A Bird Calendar for Northern India

Douglas Dewar

Alpha Editions

Copyright © 2018

ISBN : 9789352978564

Design and Setting By
Alpha Editions
email - alphaedis@gmail.com

All rights reserved. No part of this publication may be reproduced, distributed, or transmitted in any form or by means, including photocopying, recording, or other electronic or mechanical methods, without the prior written permission of the publisher.

The views and characters expressed in the book are of the author and his/her imagination and do not represent the views of the Publisher.

Contents

JANUARY	- 1 -
FEBRUARY	- 10 -
MARCH	- 18 -
APRIL	- 32 -
MAY	- 41 -
JUNE	- 54 -
JULY	- 61 -
AUGUST	- 71 -
SEPTEMBER	- 79 -
OCTOBER	- 86 -
NOVEMBER	- 93 -
DECEMBER	- 99 -
GLOSSARY	- 104 -

JANUARY

Up—let us to the fields away,
And breathe the fresh and balmy air.

MARY HOWITT.

Take nine-and-twenty sunny, bracing English May days, steal from March as many still, starry nights, to these add two rainy mornings and evenings, and the product will resemble a typical Indian January. This is the coolest month in the year, a month when the climate is invigorating and the sunshine temperate. But even in January the sun's rays have sufficient power to cause the thermometer to register 70° in the shade at noon, save on an occasional cloudy day.

Sunset is marked by a sudden fall of temperature. The village smoke then hangs a few feet above the earth like a blue-grey diaphanous cloud.

The cold increases throughout the hours of darkness. In the Punjab hoar-frosts form daily; and in the milder United Provinces the temperature often falls sufficiently to allow of the formation of thin sheets of ice. Towards dawn mists collect which are not dispersed until the sun has shone upon them for several hours. The vultures await the dissipation of these vapours before they ascend to the upper air, there to soar on outstretched wings and scan the earth for food.

On New Year's Day the wheat, the barley, the gram, and the other Spring crops are well above the ground, and, ere January has given place to February, the emerald shoots of the corn attain a height of fully sixteen inches. On these the geese levy toll.

Light showers usually fall in January. These are very welcome to the agriculturalist because they impart vigour to the young crops. In the seasons when the earth is not blessed with the refreshing winter rain men and oxen are kept busy irrigating the fields. The cutting and the pressing of the sugar-cane employ thousands of husbandmen and their cattle. In almost every village little sugar-cane presses are being worked by oxen from sunrise to sunset. At night-time the country-side is illuminated by the flames of the *megas* burned by the rustic sugar-boilers.

January is the month in which the avian population attains its maximum. Geese, ducks, teal, pelicans, cormorants, snake-birds and ospreys abound in the rivers and *jhils*; the marshes and swamps are the resort of millions of snipe and other waders; the fields and groves swarm with flycatchers, chats, starlings, warblers, finches, birds of prey and the other migrants which in winter visit the plains from the Himalayas and the country beyond.

The bracing climate of the Punjab attracts some cold-loving species for which the milder United Provinces have no charms. Conspicuous among these are rooks, ravens and jackdaws. On the other hand, frosts drive away from the Land of the Five Rivers certain of the feathered folk which do not leave the United Provinces or Bengal: to wit, the purple sunbird, the bee-eater and, to a large extent, the king-crow.

The activity of the feathered folk is not at its height in January. Birds are warm-blooded creatures and they love not the cold. Comparatively few of them are in song, and still fewer nest, at this season.

Song and sound are expressions of energy. Birds have more vitality, more life in them than has any other class of organism. They are, therefore, the most noisy of beings.

Many of the calls of birds are purposeful, being used to express pleasure or anger, or to apprise members of a flock of one another's presence. Others appear to serve no useful end. These are simply the outpourings of superfluous energy, the expressions of the supreme happiness that perfect health engenders. Since the vigour of birds is greatest at the nesting season, it follows that that is the time when they are most vociferous. Some birds sing only at the breeding season, while others emit their cries at all times. Hence the avian choir in India, as in all other countries, is composed of two sets of vocalists—those who perform throughout the year, "the musicians of all times and places," and those who join the chorus only for a few weeks or months. The calls of the former class go far to create for India its characteristic atmosphere. To enumerate all such bird calls would be wearisome.

For the purposes of this calendar it is necessary to describe only the common daily cries—the sounds that at all times and all seasons form the basis of the avian chorus.

From early dawn till nightfall the welkin rings with the harsh caw of the house-crow, the deeper note of the black crow or corby, the tinkling music of the bulbuls, the cheery *keky, keky, kek, kek* ... *chur, chur, kok, kok, kok* of the myna, the monotonous *cuckoo-coo-coo* of the spotted dove (*Turtur suratensis*), the soft subdued *cuk-cuk-coo-coo-coo* of the little brown dove (*T. cambayensis*), the mechanical *ku-ku—ku* of the ring-dove (*T. risorius*), the loud penetrating shrieks of the green parrot, the trumpet-like calls of the saras crane, the high-pitched *did-he-do-it* of the red-wattled lapwing, the wailing trill *chee-hee-hee-hee hee—hee* of the kite, the hard grating notes and the metallic *coch-lee, coch-lee* of the tree-pie; the sharp *towee, towee, towee* of the tailor-bird, the soft melodious cheeping calls of the flocks of little white-eyes, the *chit, chit, chitter* of the sparrow, the screaming cries of the golden-backed woodpecker, the screams and the trills of the white-breasted kingfisher, the curious harsh clamour of the cuckoo-shrike, and, last but by no means least, the sweet and cheerful whistling refrain of the fan-tail flycatcher, which at frequent intervals emanates from a tree in the garden or the mango tope. Nor is the bird choir altogether hushed during the hours of darkness. Throughout the year, more especially on moonlit nights, the shrieking *kucha, kwachee, kwachee, kwachee, kwachee* of the little spotted owlet disturbs the silences of the moon. Few nights pass on which the dusky horned owl fails to utter his grunting hoot, or the jungle owlet to emit his curious but not unpleasant *turtuck, turtuck, turtuck, turtuck, turtuck, tukatu, chatuckatuckatuck*.

The above are the commonest of the bird calls heard throughout the year. They form the basis of the avian melody in India. This melody is reinforced from time to time by the songs of those birds that may be termed the seasonal choristers. It is the presence or absence of the voices of these latter which imparts distinctive features to the minstrelsy of every month of the year.

In January the sprightly little metallic purple sunbird pours forth, from almost every tree or bush, his powerful song, which, were it a little less sharp, might easily be mistaken for that of a canary.

From every mango tope emanates a loud "Think of me ... Never to be." This is the call of the grey-headed flycatcher (*Culicicapa ceylonensis*), a bird that visits the plains of northern India every winter. In summer it retires to the Himalayas for nesting purposes. Still more melodious is the call of the wood-shrike, which is frequently heard at this season, and indeed during the greater part of the year.

Every now and again the green barbet emits his curious chuckling laugh, followed by a monotonous *kutur, kutur, kuturuk*. At rare intervals his cousin, the coppersmith, utters a soft *wow* and thereby reminds us that he is in the land of the living. These two species, more especially the latter, seem to dislike the cold weather. They revel in the heat; it is when the thermometer stands at something over 100° in the shade that they feel like giants refreshed, and repeat their loud calls with wearying insistence throughout the hours of daylight.

The nuthatches begin to tune up in January. They sing with more cheer than harmony, their love-song being a sharp penetrating *tee-tee-tee-tee-tee*.

The hoopoe reminds us of his presence by an occasional soft *uk-uk-uk*. His breeding season, like that of the nuthatch, is about to begin.

The magpie-robin or *dhayal*, who for months past has uttered no sound, save a scolding note when occasion demanded, now begins to make melody. His January song, however, is harsh and crude, and not such as to lead one to expect the rich deep-toned music that will compel admiration in April, May and June.

Towards the end of the month the fluty call of the koel, another hot-weather chorister, may be heard in the eastern portions of northern India.

Most of the cock sunbirds cast off their workaday plumage and assumed their splendid metallic purple wedding garment in November and December, a few, however, do not attain their full glory until January. By the end of the month it is difficult to find a cock that is not bravely attired from head to tail in iridescent purple.

Comparatively few birds build their nests in January. Needless to state, doves' nests containing eggs may be found at this season as at all other seasons. It is no exaggeration to assert that some pairs of doves rear up seven or eight broods in the course of the year. The consequence is that, notwithstanding the fact that the full clutch consists of but two eggs, doves share with crows, mynas, sparrows and green parrots the distinction of being the most successful birds in India.

The nest of the dove is a subject over which most ornithologists have waxed sarcastic. One writer compares the structure to a bundle of spillikins. Another says, "Upset a box of matches in a bush and you will have produced a very fair imitation of a dove's nursery!" According to a third, the best way to make an imitation dove's nest is to take four slender twigs, lay two of them on a branch and then place the remaining two crosswise on top of the first pair. For all this, the dove's nest is a wonderful structure; it is a lesson in how to make a little go a long way. Doves seem to place their nurseries haphazard on the first branch or ledge they come across after the spirit has moved them to build. The nest appears to be built solely on considerations of hygiene. Ample light and air are a *sine qua non*; concealment appears to be a matter of no importance.

In India winter is the time of year at which the larger birds of prey, both diurnal and nocturnal, rear up their broods. Throughout January the white-backed vultures are occupied in parental duties. The breeding season of these birds begins in October or November and ends in February or March. The nest, which is placed high up in a lofty tree, is a large platform composed of twigs which the birds themselves break off from the growing tree. Much amusement may be derived from watching the struggles of a white-backed vulture when severing a tough branch. Its wing-flapping and its tugging cause a great commotion in the tree. The boughs used by vultures for their nests are mostly covered with green leaves. These last wither soon after the branch has been plucked, so that, after the first few days of its existence, the nest looks like a great ball of dead leaves caught in a tree.

The nurseries of birds of prey can be described neither as picturesque nor as triumphs of architecture, but they have the great merit of being easy to see. January is the month in which to look for the eyries of Bonelli's eagles (*Hieraetus fasciatus*); not that the search is likely to be successful. The high cliffs of the Jumna and the Chambal in the Etawah district are the only places where the nests of this fine eagle have been recorded in the United Provinces. Mr. A. J. Currie has found the nest on two occasions in a mango tree in a tope at Lahore. In each case the eyrie was a flat platform of sticks about twice the size of a kite's nest. The ground beneath the eyrie was littered with fowls' feathers and pellets of skin, fur and bone. Most of these pellets contained squirrels' skulls; and Mr. Currie actually saw one of the parent birds fly to the nest with a squirrel in its talons.

Bonelli's eagle, when sailing through the air, may be recognised by the long, hawk-like wings and tail, the pale body and dark brown wings. It soars in circles, beating its pinions only occasionally.

The majority of the tawny eagles (*Aquila vindhiana*) build their nests in December. By the middle of January many of the eggs have yielded nestlings which are covered with white down. In size and appearance the tawny eagle is not unlike a kite. The shape of the tail, however, enables the observer to distinguish between the two species at a glance. The tail of the kite is long and forked, while that of the eagle is short and rounded at the extremity. The Pallas's fishing-eagles (*Haliaetus leucoryphus*) are likewise busy feeding their young. These fine birds are readily identified by the broad white band in the tail. Their loud resonant but unmelodious calls make it possible to recognise them when they are too far off for the white tail band to be distinguished.

This species is called a fishing-eagle; but it does not indulge much in the piscatorial art. It prefers to obtain its food by robbing ospreys, kites, marsh-harriers and other birds weaker than itself. So bold is it that it frequently swoops down and carries off a dead or wounded duck shot by the sportsman. Another raptorial bird of which the nest is likely to be found in January is the *Turumti* or red-headed merlin (*Aesalon chicquera*). The nesting season of this ferocious pigmy extends from January to May, reaching its height during March in the United Provinces and during April in the Punjab.

As a general rule birds begin nesting operations in the Punjab from fifteen to thirty days later than in the United Provinces. Unless expressly stated the times mentioned in this calendar relate to the United Provinces. The nest of the red-headed merlin is a compact circular platform, about twelve inches in diameter, placed in a fork near the top of a tree.

The attention of the observer is often drawn to the nests of this species, as also to those of other small birds of prey and of the kite, by the squabbles that occur between them and the crows. Both species of crow seem to take great delight in teasing raptorial birds. Sometimes two or three of the *corvi* act as if they had formed a league for the prevention of nest-building on the part of white-eyed buzzards, kites, shikras and other of the lesser birds of prey. The *modus operandi* of the league is for two or more of its members to hie themselves to the tree in which the victim is building its nest, take up positions near that structure and begin to caw derisively. This invariably provokes the owners of the nest to attack the black villains, who do not resist, but take to their wings. The angry, swearing builders follow in hot pursuit for a short distance and then fly back to the nest. After a few minutes the crows return. Then the performance is repeated; and so on, almost *ad infinitum*. The result is that many pairs of birds of prey take three weeks or longer to construct a nest which they could have completed within a week had they been unmolested.

Most of the larger owls are now building nests or sitting on eggs; a few are seeking food for their offspring. As owls work on silent wing at night, they escape the attentions of the crows and the notice of the average human being. The nocturnal birds of prey of which nests are likely to be found in January are the brown fish-owl (*Ketupa ceylonensis*) and the rock and the dusky horned-owls (*Bubo bengalensis* and *B. coromandus*). The dusky horned-owl builds a stick nest in a tree, the rock horned-owl lays its eggs on the bare ground or on the ledge of a cliff, while the brown fish-owl makes a nest among the branches or in a hollow in the trunk of a tree or on the ledge of a cliff.

In the Punjab the ravens, which in many respects ape the manners of birds of prey, are now nesting. A raven's nest is a compact collection of twigs. It is usually placed in an isolated tree of no great size.

The Indian raven has not the austere habits of its English brother. It is fond of the society of its fellows. The range of this fine bird in the plains of India is confined to the North-West Frontier Province Sind, and the Punjab.

An occasional pair of kites may be seen at work nest-building during the present month.

Some of the sand-martins (*Cotyle sinensis*), likewise, are engaged in family duties. The river bank in which a colony of these birds is nesting is the scene of much animation. The bank is riddled with holes, each of which, being the entrance to a martin's nest, is visited a score of times an hour by the parent birds, bringing insects captured while flying over the water.

Some species of munia breed at this time of the year. The red munia, or amadavat, or *lal* (*Estrelda amandava*) is, next to the paroquet, the bird most commonly caged in India. This little exquisite is considerably smaller than a sparrow. Its bill is bright crimson, and there is some red or crimson in the plumage—more in the cock than in the hen, and most in both sexes at the breeding season. The remainder of the plumage is brown, but is everywhere heavily spotted with white. In a state of nature these birds affect long grass, for they feed largely, if not entirely, on grass seed. The cock has a sweet voice, which, although feeble, is sufficiently loud to be heard at some distance and is frequently uttered.

The nest of the amadavat is large for the size of the bird, being a loosely-woven cup, which is egg-shaped and has a hole at or near the narrow end. It is composed of fine grass stems and is often lined with soft material. It is usually placed in the middle of a bush, sometimes in a tussock of grass. From six to fourteen eggs are laid. These are white in colour. This species appears to breed twice in the year—from October to February and again from June to August.

The white-throated munia (*Uroloncha malabarica*) is a dull brown bird, with a white patch above the tail. Its throat is yellowish white. The old name for the bird—the plain brown munia—seems more appropriate than that with which the species has since been saddled by Blanford. The nest of this little bird is more loosely put together and more globular than that of the amadavat. It is usually placed low down in a thorny bush. The number of eggs laid varies from six to fifteen. These, like those of the red munia, are white. June seems to be the only month in the year in which the eggs of this species have not been found. In the United Provinces more nests containing eggs are discovered in January than in any other month.

Occasionally in January a pair of hoopoes (*Upupa indica*) steals a march on its brethren by selecting a nesting site and laying eggs. Hoopoes nest in holes in trees or buildings. The aperture to the nest cavity is invariably small. The hen hoopoe alone incubates, and as, when once she has begun to sit, she rarely, if ever, leaves the nest till the eggs are hatched, the cock has to bring food to her. But, to describe the nesting operations of the hoopoe in January is like talking of cricket in April. It is in February and March that the hoopoes nest in their millions, and call softly, from morn till eve, *uk-uk-uk*.

Of the other birds which nest later in the season mention must be made in the calendar for the present month of the Indian cliff-swallow (*Hirundo fluvicola*) and the blue rock-pigeon (*Columba intermedia*), because their nests are sometimes seen in January.

FEBRUARY

> There's perfume upon every wind,
> Music in every tree,
> Dews for the moisture-loving flowers,
> Sweets for the sucking-bee.
>
> N. P. WILLIS.

Even as January in northern India may be compared to a month made up of English May days and March nights, so may the Indian February be likened to a halcyon month composed of sparkling, sun-steeped June days and cool starlit April nights.

February is the most pleasant month of the whole year in both the Punjab and the United Provinces; even November must yield the palm to it. The climate is perfect. The nights and early mornings are cool and invigorating; the remainder of each day is pleasantly warm; the sun's rays, although gaining strength day by day, do not become uncomfortably hot save in the extreme south of the United Provinces. The night mists, so characteristic of December and January, are almost unknown in February, and the light dews that form during the hours of darkness disappear shortly after sunrise.

The Indian countryside is now good to look upon; it possesses all the beauties of the landscape of July; save the sunsets. The soft emerald hue of the young wheat and barley is rendered more vivid by contrast with the deep rich green of the mango trees. Into the earth's verdant carpet is worked a gay pattern of white poppies, purple linseed blooms, blue and pink gram flowers, and yellow blossoms of mimosa, mustard and _arhar_. Towards the end of the month the silk-cotton trees (_Bombax malabarica_) begin to put forth their great red flowers, but not until March does each look like a great scarlet nosegay.

The patches of sugar-cane grow smaller day by day, and in nearly every village the little presses are at work from morn till eve.

From the guava groves issue the rattle of tin pots and the shouts of the boys told off to protect the ripening fruit from the attacks of crows, parrots and other feathered marauders. Nor do these sounds terminate at night-fall; indeed they become louder after dark, for it is then that the flying-foxes come forth and work sad havoc among fruit of all descriptions.

The fowls of the air are more vivacious than they were in January. The bulbuls tinkle more blithely, the purple sunbirds sing more lustily; the *kutur, kutur, kuturuk* of the green barbets is uttered more vociferously; the nuthatches now put their whole soul into their loud, sharp *tee-tee-tee-tee*, the hoopoes call *uk-uk-uk* more vigorously.

The coppersmiths (*Xantholaema haematocephala*) begin to hammer on their anvils—*tonk-tonk-tonk-tonk*, softly and spasmodically in the early days of the month, but with greater frequency and intensity as the days pass. The brain-fever bird (*Hierococcyx varius*) announces his arrival in the United Provinces by uttering an occasional "brain-fever." As the month draws to its close his utterances become more frequent. But his time is not yet. He merely gives us in February a foretaste of what is to come.

The *tew* of the black-headed oriole (*Oriolus melanocephalus*), which is the only note uttered by the bird in the colder months, is occasionally replaced in February by the summer call of the species—a liquid, musical *peeho*. In the latter half of the month the Indian robin (*Thamnobia cambayensis*) begins to find his voice. Although not the peer of his English cousin, he is no mean singer. At this time of year, however, his notes are harsh. He is merely "getting into form."

The feeble, but sweet, song of the crested lark or *Chandul* is one of the features of February. The Indian skylark likewise may now be heard singing at Heaven's gate in places where there are large tracts of uncultivated land. As in January so in February the joyous "Think of me ... Never to be" of the grey-headed flycatcher emanates from every tope.

By the middle of the month the pied wagtails and pied bush chats are in full song. Their melodies, though of small volume, are very sweet.

The large grey shrikes add the clamour of their courtship to the avian chorus.

Large numbers of doves, vultures, eagles, red-headed merlins, martins and munias—birds whose nests were described in January—are still busy feeding their young.

The majority of the brown fish-owls (*Ketupa ceylonensis*) and rock horned-owls (*Bubo bengalensis*) are sitting; a few of them are feeding young birds. The dusky horned-owls (*B. coromandus*) have either finished breeding or are tending nestlings. In addition to the nests of the above-mentioned owls those of the collared scops owl (*Scops bakkamaena*) and the mottled wood-owl (*Syrnium ocellatum*) are likely to be found at this season of the year. The scops is a small owl with aigrettes or "horns," the wood-owl is a large bird without aigrettes.

Both nest in holes in trees and lay white eggs after the manner of their kind. The scops owl breeds from January till April, while February and March are the months in which to look for the eggs of the wood-owl.

In the western districts of the United Provinces the Indian cliff-swallows (*Hirundo fluvicola*) are beginning to construct their curious nests. Here and there a pair of blue rock-pigeons (*Colombia intermedia*) is busy with eggs or young ones. In the Punjab the ravens are likewise employed.

The nesting season of the hoopoe has now fairly commenced. Courtship is the order of the day. The display of this beautiful species is not at all elaborate. The bird that "shows off" merely runs along the ground with corona fully expanded. Mating hoopoes, however, perform strange antics in the air; they twist and turn and double, just as a flycatcher does when chasing a fleet insect. Both the hoopoe and the roller are veritable aerial acrobats. By the end of the month all but a few of the hoopoes have begun to nest; most of them have eggs, while the early birds, described in January as stealing a march on their brethren, are feeding their offspring. The 6th February is the earliest date on which the writer has observed a hoopoe carrying food to the nest; that was at Ghazipur.

March and April are the months in which the majority of coppersmiths or crimson-breasted barbets rear up their families. Some, however, are already working at their nests. The eggs are hatched in a cavity in a tree—a cavity made by means of the bird's bill. Both sexes take part in nest construction. A neatly-cut circular hole, about the size of a rupee, on the lower surface or the side of a branch is assuredly the entrance to the nest of a coppersmith, a green barbet, or a woodpecker.

As the month draws to its close many a pair of nuthatches (*Sitta castaneiventris*) may be observed seeking for a hollow in which to nestle. The site selected is usually a small hole in the trunk of a mango tree that has weathered many monsoons. The birds reduce the orifice of the cavity to a very small size by plastering up the greater part of it with mud. Hence the nest of the nuthatch, unless discovered when in course of construction, is difficult to locate.

All the cock sunbirds (*Arachnechthra asiatica*) are now in the full glory of their nuptial plumage. Here and there an energetic little hen is busily constructing her wonderful pendent nest. Great is the variety of building material used by the sunbird. Fibres, slender roots, pliable stems, pieces of decayed wood, lichen, thorns and even paper, cotton and rags, are pressed into service. All are held together by cobweb, which is the favourite cement of bird masons. The general shape of the nest is that of a pear. Its contour is often irregular, because some of the materials hang loosely from the outer surface.

The nursery is attached by means of cobweb to the beam or branch from which it hangs. It is cosily lined with cotton or other soft material. The hen, who alone builds the nest and incubates the eggs, enters and leaves the chamber by a hole at one side. This is protected by a little penthouse. The door serves also as window. The hen rests her chin on the lower part of this while she is incubating her eggs, and thus is able, as she sits, to see what is going on in the great world without. She displays little fear of man and takes no pains to conceal her nest, which is often built in the verandah of an inhabited bungalow.

As the month nears its end the big black crows (*Corvus macrorhynchus*) begin to construct their nests. The site selected is usually a forked branch of a large tree. The nest is a clumsy platform of sticks with a slight depression, lined by human or horse hair or other soft material, for the reception of the eggs. Both sexes take part in incubation. From the time the first egg is laid until the young are big enough to leave the nest this is very rarely left unguarded. When one parent is away the other remains sitting on the eggs, or, after the young have hatched out, on the edge of the nest. Crows are confirmed egg-stealers and nestling-lifters, and, knowing the guile that is in their own hearts, keep a careful watch over their offspring.

The kites (*Milvus govinda*) are likewise busy at their nurseries. At this season of the year they are noisier than usual, which is saying a great deal. They not only utter unceasingly their shrill *chee-hee-hee-hee*, but engage in many a squabble with the crows.

The nest of the kite, like that of the corby, is an untidy mass of sticks and twigs placed conspicuously in a lofty tree. Dozens of these nests are to be seen in every Indian cantonment in February and March. Why the crows and the kites should prefer the trees in a cantonment to those in the town or surrounding country has yet to be discovered.

Mention has already been made of the fact that January is the month in which the majority of the tawny eagles nest; not a few, however, defer operations till February. Hume states that, of the 159 eggs of this species of which he has a record, 38 were taken in December, 83 in January and 28 in February.

The nesting season of the white-backed vulture is drawing to a close. On the other hand, that of the black or Pondicherry vulture (*Otogyps calvus*) is beginning. This species may be readily distinguished from the other vultures, by its large size, its white thighs and the red wattles that hang down from the sides of the head like drooping ears.

The nest of this bird is a massive platform of sticks, large enough to accommodate two or three men. Hume once demolished one of these vulturine nurseries and found that it weighed over eight maunds, that is to say about six hundredweight. This vulture usually builds its nest in a lofty *pipal* tree, but in localities devoid of tall trees the platform is placed on the top of a bush.

February marks the beginning of the nesting season of the handsome pied kingfisher (*Ceryle rudis*). This is the familiar, black-and-white bird that fishes by hovering kestrel-like on rapidly-vibrating wings and then dropping from a height of some twenty feet into the water below; it is a bird greatly addicted to goldfish and makes sad havoc of these where they are exposed in ornamental ponds. The nest of the pied kingfisher is a circular tunnel or burrow, more than a yard in length, excavated in a river bank. The burrow, which is dug out by the bird, is about three inches in diameter and terminates in a larger chamber in which the eggs are laid.

Another spotted black-and-white bird which now begins nesting operations is the yellow-fronted pied woodpecker (*Liopicus mahrattensis*)—a species only a little less common than the beautiful golden-backed woodpecker. Like all the Picidae this bird nests in the trunk or a branch of a tree. Selecting a part of a tree which is decayed—sometimes a portion of the bole quite close to the ground—the woodpecker hews out with its chisel-like beak a neat circular tunnel leading to the cavity in the decayed wood in which the eggs will be deposited. The tap, tap, tap of the bill as it cuts into the wood serves to guide the observer to the spot where the woodpecker, with legs apart and tail addressed to the tree, is at work. In the same way a barbet's nest, while under construction, may be located with ease. A woodpecker when excavating its nest will often allow a human being to approach sufficiently close to witness it throw over its shoulder the chips of wood it has cut away with its bill.

In the United Provinces many of the ashy-crowned finch-larks (*Pyrrhulauda grisea*) build their nests during February. In the Punjab they breed later; April and May being the months in which their eggs are most often found in that province. These curious squat-figured little birds are rendered easy of recognition by the unusual scheme of colouring displayed by the cock—his upper parts are earthy grey and his lower plumage is black.

The habit of the finch-lark is to soar to a little height and then drop to the ground, with wings closed, singing as it descends. It invariably affects open plains. There are very few tracts of treeless land in India which are not tenanted by finch-larks. The nest is a mere pad of grass and feathers placed on the ground in a tussock of grass, beside a clod of earth, or in a depression, such as a hoof-print. The most expeditious way of finding nests of these birds in places where they are abundant is to walk with a line of beaters over a tract of fallow land and mark carefully the spots from which the birds rise.

With February the nesting season of the barn-owls (*Strix flammea*) begins in the United Provinces, where their eggs have been taken as early as the 17th.

Towards the end of the month the white-browed fantail flycatchers (*Rhipidura albifrontata*) begin to nest. The loud and cheerful song of this little feathered exquisite is a tune of six or seven notes that ascend and descend the musical scale. It is one of the most familiar of the sounds that gladden the Indian countryside. The broad white eyebrow and the manner in which, with drooping wings and tail spread into a fan, this flycatcher waltzes and pirouettes among the branches of a tree render it unmistakable. The nest is a dainty little cup, covered with cobweb, attached to one of the lower boughs of a tree. So small is the nursery that sometimes the incubating bird looks as though it were sitting across a branch. This species appears to rear two broods every year. The first comes into existence in March or late February in the United Provinces and five or six weeks later in the Punjab; the second brood emerges during the monsoon.

The white-eyed buzzards—weakest of all the birds of prey—begin to pair towards the end of the month. At this season they frequently rise high above the earth and soar, emitting plaintive cries.

The handsome, but destructive, green parrots are now seeking, or making, cavities in trees or buildings in which to deposit their white eggs.

The breeding season for the alexandrine (*Palaeornis eupatrius*) and the rose-ringed paroquet (*P. torquatus*) begins at the end of January or early in February. March is the month in which most eggs are taken.

In April and May the bird-catchers go round and collect the nestlings in order to sell them at four annas apiece. Green parrots are the most popular cage birds in India. Destructive though they be and a scourge to the husbandman, one cannot but pity the luckless captives doomed to spend practically the whole of their existence in small iron cages, which, when exposed to the sun in the hot weather, as they often are, must be veritable infernos.

The courtship of a pair of green parrots is as amusing to watch as that of any 'Arry and 'Arriet. Not possessing hats the amorous birds are unable to exchange them, but otherwise their actions are quite coster-like. The female twists herself into all manner of ridiculous postures and utters low twittering notes. The cock sits at her side and admires. Every now and then he shows his appreciation of her antics by tickling her head with his beak or by joining his bill to hers.

Both the grey shrike and the wood-shrike begin nesting operations in February. As, however, most of their nests are likely to be found later in the year they are dealt with in the calendar for March.

MARCH

> And all the jungle laughed with nesting songs,
> And all the thickets rustled with small life
> Of lizard, bee, beetle, and creeping things
> Pleased at the spring time. In the mango sprays
> The sun-birds flashed; alone at his green forge
> Toiled the loud coppersmith; . . .
>
> ARNOLD. *The Light of Asia.*

In March the climate of the plains of the United Provinces varies from place to place. In the western sub-Himalayan tracts, as in the Punjab, the weather still leaves little to be desired. The sun indeed is powerful; towards the end of the month the maximum shade temperature exceeds 80°, but the nights and early mornings are delightfully cool. In all the remaining parts of the United Provinces, except the extreme south, temperate weather prevails until nearly the end of the month. In the last days the noonday heat becomes so great that many persons close their bungalows for several hours daily to keep them cool, the outer temperature rising to ninety in the shade. At night, however, the temperature drops to 65°. In the extreme south of the Province the hot weather sets in by the middle of March. The sky assumes a brazen aspect and, at midday, the country is swept by westerly winds which seem to come from a titanic blast furnace.

The spring crops grow more golden day by day. The mustard is the first to ripen. The earlier-sown fields are harvested in March in the eastern and southern parts of the country. The spring cereals are cut by hand sickles, the grain is then husked by the tramping of cattle, and, lastly, the chaff is separated from the grain on the threshing floor, the hot burning wind often acting as a natural winnowing fan.

The air is heavily scented with the inconspicuous inflorescences of the mangos (*Mangifera indica*). The pipals (*Ficus religiosa*) are shedding their leaves; the *sheshams* (*Dalbergia sissoo*) are assuming their emerald spring foliage.

The garden, the jungle and the forest are beautified by the gorgeous reds of the flowers of the silk-cotton tree (*Bombax malabarica*), the Indian coral tree (*Erythrina indica*) and the flame-of-the-forest (*Butea frondosa*). The sub-Himalayan forests become yellow-tinted owing to the fading of the leaves of the *sal* (*Shorea robusta*), many of which are shed in March. The *sal*, however, is never entirely leafless; the young foliage appears as the old drops off; while this change is taking place the minute pale yellow flowers open out.

The familiar yellow wasps, which have been hibernating during the cold weather, emerge from their hiding-places and begin to construct their umbrella-shaped nests or combs, which look as if they were made of rice-paper.

March is a month of great activity for the birds. Those that constituted the avian chorus of February continue to sing, and to their voices are now added those of many other minstrels. Chief of these is the pied singer of Ind—the magpie-robin or *dhayal*—whose song is as beautiful as that of the English robin at his best. From the housetops the brown rock-chat begins to pour forth his exceedingly sweet lay. The Indian robin is in full song. The little golden ioras, hidden away amid dense foliage, utter their many joyful sounds. The brain-fever bird grows more vociferous day by day. The crow-pheasants, which have been comparatively silent during the colder months of the year, now begin to utter their low sonorous *whoot, whoot, whoot*, which is heard chiefly at dawn.

Everywhere the birds are joyful and noisy; nowhere more so than at the silk-cotton and the coral trees. These, although botanically very different, display many features in common. They begin to lose their leaves soon after the monsoon is over, and are leafless by the end of the winter. In the early spring, while the tree is still devoid of foliage, huge scarlet, crimson or yellow flowers emerge from every branch. Each flower is plentifully supplied with honey; it is a flowing bowl of which all are invited to partake, and hundreds of thousands of birds accept the invitation with right good-will. The scene at each of these trees, when in full flower, baffles description.

Scores of birds forgather there—rosy starlings, mynas, babblers, bulbuls, king-crows, tree-pies, green parrots, sunbirds and crows.

These all drink riotously and revel so loudly that the sound may be heard at a distance of half a mile or more. Even before the sun has risen and begun to dispel the pleasant coolness of the night the drinking begins. It continues throughout the hours of daylight. Towards midday, when the west wind blows very hot, it flags somewhat, but even when the temperature is nearer 100° than 90° some avian brawlers are present. As soon as the first touch of the afternoon coolness is felt the clamour acquires fresh vigour and does not cease until the sun has set in a dusty haze, and the spotted owlets have emerged and begun to cackle and call as is their wont.

These last are by no means the only birds that hold concert parties during the hours of darkness. In open country the jungle owlet and the dusky-horned owl call at intervals, and the Indian nightjar (*Caprimulgus asiaticus*) imitates the sound of a stone skimming over ice. In the forest tracts Franklin's and Horsfield's nightjars make the welkin ring. Scarce has the sun disappeared below the horizon when the former issues forth and utters its harsh *tweet*. Horsfield's nightjar emerges a few minutes later, and, for some hours after dusk and for several before dawn, it utters incessantly its loud monotonous *chuck, chuck, chuck, chuck, chuck*, which has been aptly compared to the sound made by striking a plank sharply with a hammer.

March is the month in which the majority of the shrikes or butcher-birds go a-courting. There is no false modesty about butcher-birds. They are not ashamed to introduce their unmelodious calls into the avian chorus. But they are mild offenders in comparison with the king-crows (*Dicrurus ater*) and the rollers (*Coracias indica*).

The little black king-crows are at all seasons noisy and vivacious: from the end of February until the rains have set in they are positively uproarious. Two or three of them love to sit on a telegraph wire, or a bare branch of a tree, and hold a concert. The first performer draws itself up to its full height and then gives vent to harsh cries. Before it has had time to deliver itself of all it has to sing, an impatient neighbour joins in and tries to shout it down. The concert may last for half an hour or longer; the scene is shifted from time to time as the participants become too excited to sit still. The king-crows so engaged appear to be selecting their mates;

nevertheless nest-construction does not begin before the end of April.

Some human beings may fail to notice the courtship of the king-crow, but none can be so deaf and blind as to miss the love-making of the gorgeous roller or blue jay. Has not everyone marvelled at the hoarse cries and rasping screams which emanate from these birds as they fling themselves into the air and ascend and descend as though they were being tossed about by unseen hands?

Their wonderful aerial performances go on continually in the hours of daylight throughout the months of March and April; at this season the birds, beautiful although they be, are a veritable nuisance, and most people gratefully welcome the comparative quiet that supervenes after the eggs have been laid. The madness of the March hare is mild compared with that of the March roller. It is difficult to realise that the harsh and angry-sounding cries of these birds denote, not rage, but joy.

The great exodus of the winter visitors from the plains of India begins in March. It continues until mid-May, by which time the last of the migratory birds will have reached its distant breeding ground.

This exodus is usually preceded by the gathering into flocks of the rose-coloured starlings and the corn-buntings. Large noisy congregations of these birds are a striking feature of February in Bombay, of March in the United Provinces, and of April in the Punjab.

Rose-coloured starlings spend most of their lives in the plains of India, going to Asia Minor for a few months each summer for nesting purposes. In the autumn they spread themselves over the greater part of Hindustan, most abundantly in the Deccan.

In the third or fourth week of February the rosy starlings of Bombay begin to form flocks. These make merry among the flowers of the coral tree, which appear first in South India, and last in the Punjab. The noisy flocks journey northwards in a leisurely manner, timing their arrival at each place simultaneously with the flowering of the coral trees. They feed on the nectar provided by these flowers and those of the silk-cotton tree. They also take toll of the ripening corn and of the mulberries which are now in season. Thus the rosy starlings reach Allahabad about the second week in March, and Lahore some fifteen days later.

The head, neck, breast, wings and tail of the rosy starling are glossy black, and the remainder of the plumage is pale salmon in the hen and the young cock, and faint rose-colour in the adult cock.

Rosy starlings feed chiefly in the morning and the late afternoon. During the hottest part of the day they perch in trees and hold a concert, if such a term may be applied to a torrent of sibilant twitter.

Buntings, like rosy starlings, are social birds, and are very destructive to grain crops.

As these last are harvested the feeding area of the buntings becomes restricted, so that eventually every patch of standing crop is alive with buntings. The spring cereals ripen in the south earlier than in northern India, so that the cheerful buntings are able to perform their migratory journey by easy stages and find abundant food all along the route.

There are two species of corn-bunting—the red-headed (*Emberiza luteola*) and the black-headed (*E. melanocephala*). In both the lower plumage is bright yellow.

Among the earliest of the birds to forsake the plains of Hindustan are the grey-lag goose and the pintail duck. These leave Bengal in February, but tarry longer in the cooler parts of the country. Of the other migratory species many individuals depart in March, but the greater number remain on into April, when they are caught up in the great migratory wave that surges over the country. The destination of the majority of these migrants is Tibet or Siberia, but a few are satisfied with the cool slopes of the Himalayas as a summer resort in which to busy themselves with the sweet cares of nesting. Examples of these more local migrants are the grey-headed and the verditer flycatchers, the Indian bush-chat and, to some extent, the paradise flycatcher and the Indian oriole. The case of the oriole is interesting. All the Indian orioles (*Oriolus kundoo*) disappear from the Punjab and the United Provinces in winter. In the former province no other oriole replaces *O. kundoo*, but in the United Provinces the black-headed oriole (*O. melanocephalus*) comes to take the place of the other from October to March. When this last returns to the United Provinces in March the greater number of *melanocephalus* individuals go east, a few only remaining in the sub-Himalayan tracts of the province.

The Indian oriole is not the only species which finds the climate of the United Provinces too severe for it in winter; the koel and the paradise flycatcher likewise desert us in the coldest months. From the less temperate Punjab several species migrate in October which manage to maintain themselves in the United Provinces throughout the year: these are the purple sunbird, the little green and the blue-tailed bee-eaters, and the yellow-throated sparrow. The return of these and the other migrant species to the Punjab in March is as marked a phenomenon as is the arrival of the swallow and the cuckoo in England in spring.

The behaviour of the king-crows shows the marked effect a comparatively small difference of temperature may exert on the habits of some birds. In the United Provinces the king-crows appear to be as numerous in winter as in summer: in the Punjab they are very plentiful in summer, but rare in the cold weather; while not a single king-crow winters in the N.-W. Frontier Province.

Of the birds of which the nests were described in January and February the Pallas's fishing eagles have sent their nestlings into the world to fend for themselves.

In the case of the following birds the breeding season is fast drawing to its close:—the dusky horned-owl, the white-backed vulture, Bonelli's eagle, the tawny eagle, the brown fish-owl, the rock horned-owl, the raven, the amadavat and the white-throated munia.

The nesting season is at its height for all the other birds of which the nests have been described, namely, most species of dove, the jungle crow, the red-headed merlin, the purple sunbird, the nuthatch, the fantail flycatcher, the finch-lark, the pied woodpecker, the coppersmith, the alexandrine and the rose-ringed paroquet, the white-eyed buzzard, the collared scops and the mottled wood-owl, the kite, the black vulture and the pied kingfisher.

The sand-martins breed from October to May, consequently their nests, containing eggs or young, are frequently taken in March. Mention was made in January and February of the Indian cliff-swallow (*Hirundo fluvicola*). This species is not found in the eastern districts of the United Provinces, but it is the common swallow of the western districts. The head is dull chestnut. The back and shoulders are glistening steel-blue. The remainder of the upper plumage is brown. The lower parts are white with brown streaks, which are most apparent on the throat and upper breast. These swallows normally nest at two seasons of the year—from February till April and in July or August.

They breed in colonies. The mud nests are spherical or oval with an entrance tube from two to six inches long. The nests are invariably attached to a cliff or building, and, although isolated ones are built sometimes, they usually occur in clusters, as many as two hundred have been counted in one cluster. In such a case a section cut parallel to the surface to which the nests are attached looks like that of a huge honeycomb composed of cells four inches in diameter—cells of a kind that one could expect to be built by bees that had partaken of Mr. H. G. Wells' "food of the gods."

The beautiful white-breasted kingfisher, (*Halcyon smyrnensis*) is now busy at its nest.

This species spends most of its life in shady gardens; it feeds on insects in preference to fish. It does not invariably select a river bank in which to nest, it is quite content with a sand quarry, a bank, or the shaft of a *kachcha* well. The nest consists of a passage, some two feet in length and three inches in diameter, leading to a larger chamber in which from four to seven eggs are laid.

A pair of white-breasted kingfishers at work during the early stages of nest construction affords an interesting spectacle. Not being able to obtain a foothold on the almost perpendicular surface of the bank, the birds literally charge this in turn with fixed beak. By a succession of such attacks at one spot a hole of an appreciable size is soon formed in the soft sand. Then the birds are able to obtain a foothold and to excavate with the bill, while clinging to the edge of the hole. Every now and then they indulge in a short respite from their labours. While thus resting one of the pair will sometimes spread its wings for an instant and display the white patch; then it will close them and make a neat bow, as if to say "Is not that nice?" Its companion may remain motionless and unresponsive, or may return the compliment.

In the first days of March the bulbuls begin to breed. In 1912 the writer saw a pair of bulbuls (*Otocompsa emeria*) building a nest on the 3rd March. By the 10th the structure was complete and held the full clutch of three eggs. On that date a second nest was found containing three eggs.

In 1913 the writer first saw a bulbul's nest on the 5th March. This belonged to *Molpastes bengalensis* and contained two eggs. On the following day the full clutch of three was in the nest.

The nesting season for these birds terminates in the rains.

The common bulbuls of the plains belong to two genera—*Molpastes* and *Otocompsa*. The former is split up into a number of local species which display only small differences in appearance and interbreed freely at the places where they meet. They are known as the Madras, the Bengal, the Punjab, etc., red-vented bulbul. They are somewhat larger than sparrows. The head, which bears a short crest, and the face are black; the rest of the body, except a patch of bright red under the tail, is brown, each feather having a pale margin.

In *Otocompsa* the crest is long and rises to a sharp point which curves forward a little over the beak. The breast is white, set off by a black gorget. There is the usual red patch under the tail and a patch of the same hue on each side of the face, whence the English name for the bird—the red-whiskered bulbul.

Molpastes and *Otocompsa* have similar habits. They are feckless little birds that build cup-shaped nests in all manner of queer and exposed situations. Those that live near the habitations of Europeans nestle in low bushes in the garden, or in pot plants in the verandah. Small crotons are often selected, preferably those that do not bear a score of leaves. The sitting bulbul does not appear to mind the daily shower-bath it receives when the *mali* waters the plant. Sometimes as many as three or four pairs of bulbuls attempt to rear up families in one verandah. The word "attempt" is used advisedly, because, owing to the exposed situations in which nests are built, large numbers of eggs and young bulbuls are destroyed by boys, cats, snakes and other predaceous creatures. The average bulbul loses six broods for every one it succeeds in rearing. The eggs are pink with reddish markings.

March is the month in which to look for the nest of the Indian wren-warbler (*Prinia inornata*). *Inornata* is a very appropriate specific name for this tiny earth-brown bird, which is devoid of all kind of ornamentation. Its voice is as homely as its appearance—a harsh but plaintive *twee, twee, twee*. It weaves a nest which looks like a ragged loofah with a hole in the side. The nest is usually placed low down in a bush or in long grass. Sometimes it is attached to two or more stalks of corn. In such cases the corn is often cut before the young birds have had time to leave the nest, and then the brood perishes. This species brings up a second family in the rainy season.

The barn-owls (*Strix flammea*) are now breeding. They lay their eggs in cavities in trees, buildings or walls. In northern India the nesting season lasts from February to June. Eggs are most likely to be found in the United Provinces during the present month.

The various species of babblers or seven sisters begin to nest in March. Unlike bulbuls these birds are careful to conceal the nest. This is a slenderly-built, somewhat untidy cup, placed in a bush or tree. The eggs are a beautiful rich blue, without any markings.

The hawk-cuckoo, or brain-fever bird (*Hierococcyx varius*), to which allusion has already been made, deposits its eggs in the nests of various species of babblers. The eggs of this cuckoo are blue, but are distinguishable from those of the babbler by their larger size. It may be noted, in passing, that this cuckoo does not extend far into the Punjab.

As stated above, most of the shrikes go a-courting in March. Nest-building follows hard on courtship. In this month and in April most of the shrikes lay their eggs, but nests containing eggs or young are to be seen in May, June, July and August. Shrikes are birds of prey in miniature. Although not much larger than sparrows they are as fierce as falcons.

Their habit is to seize the quarry on the ground, after having pounced upon it from a bush or tree. Grasshoppers constitute their usual food, but they are not afraid to tackle mice or small birds.

The largest shrike is the grey species (*Lanius lahtora*). This is clothed mainly in grey; however, it has a broad black band running through the eye—the escutcheon of the butcher-bird clan. It begins nesting before the other species, and its eggs are often taken in February.

The other common species are the bay-backed (*L. vittatus*) and the rufous-backed shrike (*L. erythronotus*). These are smaller birds and have the back red. The former is distinguishable from the latter by having in the wings and tail much white, which is very conspicuous during flight.

The nest of each species is a massive cup, composed of twigs, thorns, grasses, feathers, and, usually, some pieces of rag; these last often hang down in a most untidy manner. The nest is, as a rule, placed in a babool or other thorny tree, close up against the trunk.

Three allies of the shrikes are likewise busy with their nests at this season. These are the wood-shrike, the minivet and the cuckoo-shrike. The wood-shrike (*Tephrodornis pondicerianus*) is an ashy-brown bird of the size of a sparrow with a broad white eyebrow. It frequently emits a characteristic soft, melancholy, whistling note, which Eha describes as "Be thee cheery." How impracticable are all efforts to "chain by syllables airy sounds"! The cup-like nest of this species is always carefully concealed in a tree.

Minivets are aerial exquisites. In descriptions of them superlative follows upon superlative. The cocks of most species are arrayed in scarlet and black; the hens are not a whit less brilliantly attired in yellow and sable. One species lives entirely in the plains, others visit them in the cold weather; the majority are permanent residents of the hills. The solitary denizen of the plains—the little minivet (*Pericrocotus peregrinus*)—is the least resplendent of them all. Its prevailing hue is slaty grey, but the cock has a red breast and some red on the back. The nest is a cup so small as either to be invisible from below or to present the appearance of a knot or thickening in the branch on which it is placed. Sometimes two broods are reared in the course of the year—one in March, April or May and the other during the rainy season.

The cuckoo-shrike (*Grauculus macii*) is not nearly related to the cuckoo, nor has it the parasitic habits of the latter. Its grey plumage is barred like that of the common cuckoo, hence the adjective. The cuckoo-shrike is nearly as big as a dove. It utters constantly a curious harsh call. It keeps much to the higher branches of trees in which it conceals, with great care, its saucer-like nest.

As we have seen, some coppersmiths and pied woodpeckers began nesting operations in February, but the great majority do not lay eggs until March.

The green barbet (*Thereoceryx zeylonicus*) and the golden-backed woodpecker (*Brachypternus aurantius*) are now busy excavating their nests, which are so similar to those of their respective cousins—the coppersmith and the pied woodpecker—as to require no description. It is not necessary to state that the harsh laugh, followed by the *kutur, kutur, kuturuk,* of the green barbet and the eternal *tonk, tonk, tonk* of the coppersmith are now more vehement than ever, and will continue with unabated vigour until the rains have fairly set in.

By the end of the month many of the noisy rollers have found holes in decayed trees in which the hens can lay their eggs. The vociferous nightjars likewise have laid upon the bare ground their salmon-pink eggs with strawberry-coloured markings.

The noisy spotted owlets (*Athene brama*) and the rose-ringed paroquets (*Palaeornis torquatus*) are already the happy possessors of clutches of white eggs hidden away in cavities of decayed trees or buildings.

The swifts (*Cypselus indicus*) also are busy with their nests. These are saucer-shaped structures, composed of feathers, straw and other materials made to adhere together, and to the beam or stone to which the nest is attached, by the glutinous saliva of the swifts. Deserted buildings, outhouses and verandahs of bungalows are the usual nesting sites of these birds. At this season swifts are very noisy. Throughout the day and at frequent intervals during the night they emit loud shivering screams. At sunset they hold high carnival, playing, at breakneck speed and to the accompaniment of much screaming, a game of "follow the man from Cook's."

The swifts are not the only birds engaged in rearing up young in our verandahs. Sparrows and doves are so employed, as are the wire-tailed swallows (*Hirundo smithii*). These last are steel-blue birds with red heads and white under plumage. They derive the name "wire-tailed" from the fact that the thin shafts of the outer pair of tail feathers are prolonged five inches beyond the others and look like wires. Wire-tailed swallows occasionally build in verandahs, but they prefer to attach their saucer-shaped mud nests to the arches of bridges and culverts.

With a nest in such a situation the parent birds are not obliged to go far for the mud with which the nest is made, or for the insects, caught over the surface of water, on which the offspring are fed.

The nesting season of wire-tailed swallows is a long one. According to Hume these beautiful birds breed chiefly in February and March and again in July, August and September. However, he states that he has seen eggs as early as January and as late as November. In the Himalayas he has obtained the eggs in April, May and June.

The present writer's experience does not agree with that of Hume. In Lahore, Saharanpur and Pilibhit, May and June are the months in which most nests of this species are likely to be seen. The writer has found nests with eggs or young on the following dates in the above-mentioned places: May 13th, 15th, 16th, 17th; June 6th and 28th.

The nest of June 28th was attached to a rafter of the front verandah of a bungalow at Lahore. The owner of the house stated that the swallows in question had already reared one brood that year, and that the birds in question had nested in his verandah for some years. There is no doubt that some wire-tailed swallows bring up two broods. Such would seem to breed, as Hume says, in February and March and again in July and August. But, as many nests containing eggs are found in May, some individuals appear to have one brood only, which hatches out in May or June.

Those useful but ugly fowls, the white scavenger vultures (*Neophron ginginianus*), depart from the ways of their brethren in that they nidificate in March and April instead of in January and February. The nest is an evil-smelling pile of sticks, rags and rubbish. It is placed on some building or in a tree.

The handsome brahminy kites (*Haliastur indicus*), attired in chestnut and white, are now busily occupied, either in seeking for sites or in actually building their nests, which resemble those of the common kite.

In the open plains the pipits (*Anthus rufulus*) and the crested larks (*Galerita cristata*) are keeping the nesting finch-larks company.

All three species build the same kind of nest—a cup of grass or fibres (often a deep cup in the case of the crested lark) placed on the ground in a hole or a depression, or protected by a tussock of grass or a small bush.

On the churs and sand islets in the large Indian rivers the terns are busy with their eggs, which are deposited on the bare sand. They breed in colonies. On the same islet are to be seen the eggs of the Indian river tern, the black-bellied tern, the swallow-plover, the spur-winged plover and the Indian skimmer.

The eggs of all the above species are of similar appearance, the ground colour being greenish, or buff, or the hue of stone or cream, with reddish or brownish blotches. Three is the full complement of eggs. The bare white glittering sands on which these eggs are deposited are often at noon so hot as to be painful to touch; accordingly during the daytime there is no need for the birds to sit on the eggs in order to keep them warm. Indeed, it has always been a mystery to the writer why terns' eggs laid in March in northern India do not get cooked. Mr. A. J. Currie recently came across some eggs of the black-bellied tern that had had water sprinkled over them. He is of opinion that the incubating birds treat the eggs thus in order to prevent their getting sun-baked. This theory should be borne in mind by those who visit sandbanks in March. Whether it be true or not, there is certainly no need for the adult birds to keep the eggs warm in the daytime, and they spend much of their time in wheeling gracefully overhead or in sleeping on the sand. By nightfall all the eggs are covered by parent birds, which are said to sit so closely that it is possible to catch them by means of a butterfly net. The terns, although they do not sit much on their eggs during the day, ever keep a close watch on them, so that, when a human being lands on a nest-laden sandbank, the parent birds fly round his head, uttering loud screams.

The swallow-plovers go farther. They become so excited that they flutter about on the sand, with dragging wings and limping legs, as if badly wounded. Sometimes they perform somersaults in their intense excitement. The nearer the intruder approaches their eggs the more vigorous do their antics become.

Every lover of the winged folk should make a point of visiting, late in March or early in April, an islet on which these birds nest. He will find much to interest him there. In April many of the young birds will be hatched out. A baby tern is an amusing object. It is covered with soft sand-coloured down. When a human being approaches it crouches on the sand, half burying its head in its shoulders, and remains thus perfectly motionless. If picked up it usually remains limply in the hand, so that, but for its warmth, it might be deemed lifeless. After it has been set down again on the sand, it will remain motionless until the intruder's back is turned, when it will run to the water as fast as its little legs can carry it. It swims as easily as a duck. Needless to state, the parent birds make a great noise while their young are being handled.

Birds decline to be fettered by the calendar. Many of the species which do not ordinarily nest until April or May occasionally begin operations in March, hence nests of the following species, which are dealt with next month, may occur in the present one:—the tree-pie, tailor-bird, common myna, bank-myna, brown rock-chat, brown-backed robin, pied wagtail, red-winged bush-lark, shikra, red-wattled lapwing, yellow-throated sparrow, bee-eater, blue rock-pigeon, green pigeon and grey partridge.

March the 15th marks the beginning of the close season for game birds in all the reserved forests of Northern India. This is none too soon, as some individuals begin breeding at the end of the month.

APRIL

> The breeze moves slow with thick perfume
>   From every mango grove;
> From coral tree to parrot bloom
>   The black bees questing rove,
> The koil wakes the early dawn.
>
>   WATERFIELD. *Indian Ballads*.

The fifteenth of April marks the beginning of the "official" hot weather in the United Provinces; but the elements decline to conform to the rules of man. In the eastern and southern districts hot-weather conditions are established long before mid-April, while in the sub-Himalayan belt the temperature remains sufficiently low throughout the month to permit human beings to derive some physical enjoyment from existence. In that favoured tract the nights are usually clear and cool, so that it is very pleasant to sleep outside beneath the starry canopy of the heavens.

It requires an optimist to say good things of April days, even in the sub-Himalayan tract. Fierce scorching west winds sweep over the earth, covering everything with dust. Sometimes the flying sand is so thick as to obscure the landscape, and often, after the wind has dropped, the particles remain suspended for days as a dust haze. The dust is a scourge. It is all-pervading. It enters eyes, ears, nose and mouth. To escape it is impossible. Closed doors and windows fail to keep it from entering the bungalow. The only creatures which appear to be indifferent to it are the fowls of the air. As to the heat, the non-migratory species positively revel in it. The crows and a few other birds certainly do gasp and pant when the sun is at its height, but even they, save for a short siesta at midday, are as active in April and May as schoolboys set free from a class-room. April is the month in which the spring crops are harvested. As soon as the *Holi* festival is over the cultivators issue forth in thousands, armed with sickles, and begin to reap. They are almost as active as the birds, but their activity is forced and not spontaneous; like most Anglo-Indian officials they literally earn their bread by the sweat of the brow. Thanks to their unceasing labours the countryside becomes transformed during the month; that which was a sea of smiling golden-brown wheat and barley becomes a waste of short stubble.

Nature gives some compensation for the heat and the dust in the shape of mulberries, loquats, lichis and cool luscious papitas and melons which ripen in March or April. The mango blossom becomes transfigured into fruit, which, by the end of the month, is as large as an egg, and will be ready for gathering in the latter half of May.

Many trees are in flower. The coral, the silk-cotton and the *dhak* are resplendent with red foliage. The *jhaman*, the *siris* and the *mohwa* are likewise in bloom and, ere the close of the month, the *amaltas* or Indian laburnum will put forth its bright yellow flowers in great profusion. Throughout April the air is heavy with the scent of blossoms. The *shesham*, the *sal*, the *pipal* and the *nim* are vivid with fresh foliage. But notwithstanding all this galaxy of colour, notwithstanding the brightness of the sun and the blueness of the sky, the countryside lacks the sweetness that Englishmen associate with springtime, because the majority of the trees, being evergreen, do not renew their clothing completely at this season, and the foliage is everywhere more or less obscured by the all-pervading dust.

The great avian emigration, which began in March, now reaches its height. During the warm April nights millions of birds leave the plains of India. The few geese remaining at the close of March, depart in the first days of April.

The brahminy ducks, which during the winter months were scattered in twos and threes over the lakes and rivers of Northern India, collect into flocks that migrate, one by one, to cooler climes, so that, by the end of the first week in May, the *a-onk* of these birds is no longer heard. The mallard, gadwall, widgeon, pintail, the various species of pochard and the common teal are rapidly disappearing. With April duck-shooting ends. Of the migratory species only a few shovellers and garganey teal tarry till May.

The snipe and the quail are likewise flighting towards their breeding grounds. Thus on the 1st of May the avian population of India is less by many millions than it was at the beginning of April. But the birds that remain behind more than compensate us, by their great activity, for the loss of those that have departed. There is more to interest the ornithologist in April than there was in January.

The bird chorus is now at its best. The magpie-robin is in full song. At earliest dawn he takes up a position on the topmost bough of a tree and pours forth his melody in a continuous stream. His varied notes are bright and joyous. Its voice is of wide compass and very powerful; were it a little softer in tone it would rival that of the nightingale. The magpie-robin is comparatively silent at noonday, but from sunset until dusk he sings continuously.

Throughout April the little cock sunbirds deliver themselves of their vigorous canary-like song. The bulbuls tinkle as blithely as ever. Ioras, pied wagtails, pied chats, and wood-shrikes continue to contribute their not unworthy items to the minstrelsy of the Indian countryside. The robins, having by now found their true notes, are singing sweetly and softly. The white-eyes are no longer content to utter their usual cheeping call, the cocks give vent to an exquisite warble and thereby proclaim the advent of the nesting season. The *towee, towee, towee,* of the tailor-bird, more penetrating than melodious, grows daily more vigorous, reminding us that we may now hopefully search for his nest. Among the less pleasing sounds that fill the welkin are the *tonk, tonk, tonk* of the coppersmith, the *kutur, kutur, kuturuk* of the green barbet, and the calls of the various cuckoos that summer in the plains of Northern India. The calls of these cuckoos, although frequently heard in April, are uttered more continuously in May, accordingly they are described in the calendar for that month.

The owls, of course, lift up their voices, particularly on moonlight nights. The nightjars are as vociferous as they were in March; their breeding season is now at its height.

In the hills the woods resound with the cheerful double note of the European cuckoo (*Cuculus canorus*). This bird is occasionally heard in the plains of the Punjab in April, and again from July to September, when it no longer calls in the Himalayas. This fact, coupled with the records of the presence of the European cuckoo in Central India in June and July, lends support to the theory that the birds which enliven the Himalayas in spring go south in July and winter in the Central Provinces. Cuckoos, at seasons when they are silent, are apt to be overlooked, or mistaken for shikras.

Ornithologists stationed in Central India will render a service to science if they keep a sharp look-out for European cuckoos and record the results of their observations. In this way alone can the above theory be proved or disproved.

By the middle of the month most of the rollers have settled down to domestic duties, and in consequence are less noisy than they were when courting. Their irritating grating cries are now largely replaced by harsh *tshocks* of delight, each *tshock* being accompanied by a decisive movement of the tail. The cause of these interjections expressing delight is a clutch of white eggs or a brood of young birds, hidden in a hole in a tree or a building.

April is a month in which the pulse of bird life beats very vigorously in India. He who, braving the heat, watches closely the doings of the feathered folk will be rewarded by the discovery of at least thirty different kinds of nests. Hence, it is evident that the calendar for this month, unless it is to attain very large dimensions, must be a mere catalogue of nesting species. The compiler of the calendar has to face an *embarrass de richesses*.

Of the common species that build in March and the previous months the following are likely to be found with eggs or young— the jungle crows, sunbirds, doves, pied and golden-backed woodpeckers, coppersmiths, hoopoes, common and brahminy kites, bulbuls, shrikes, little minivets, fantail flycatchers, wire-tailed swallows, paroquets, spotted owlets, swifts, scavenger vultures, red-headed merlins, skylarks, crested larks, pipits, babblers, sand-martins, cliff-swallows, nuthatches, white-eyed buzzards, kites, black vultures, pied and white-breasted kingfishers, finch-larks, Indian wren-warblers, wood-shrikes, cuckoo-shrikes, green barbets, tawny eagles, and the terns and the other birds that nest on islets in rivers. Here and there may be seen a white-backed vulture's nest containing a young bird nearly ready to fly.

Towards the middle of the month the long-tailed tree-pies (*Dendrocitta rufa*), which are nothing else than coloured crows, begin nest-building. They are to be numbered among the commonest birds in India, nevertheless their large open nests are rarely seen. The explanation of this phenomenon appears to be the fact that the nest is well concealed high up in a tree. Moreover, the pie, possessing a powerful beak which commands respect, is not obliged constantly to defend its home after the manner of small or excitable birds, and thus attract attention to it.

Fortunately for the tree-pie the kites and crows do not worry it. The shikra (*Astur badius*) and the white-eyed buzzard (*Butastur teesa*), which are now engaged in nest-building, are not so fortunate. The crows regard them as fair game, hence their nest-building season is a time of *sturm und drang*. They, in common with all diurnal birds of prey, build untidy nests in trees—mere conglomerations of sticks, devoid of any kind of architectural merit. The blue rock-pigeons (*Columba intermedia*) are busily prospecting for nesting sites. In some parts of India, especially in the Muttra and Fatehgarh districts, these birds nest chiefly in holes in wells. More often than not a stone thrown into a well in such a locality causes at least one pigeon to fly out of the well. In other places in India these birds build by preference on a ledge or a cornice inside some large building. They often breed in colonies. At Dig in Rajputana, where they are sacred in the eyes of Hindus, thousands of them nest in the fort, and, as Hume remarks, a gun fired in the moat towards evening raises a dense cloud of pigeons, "obscuring utterly the waning day and deafening one with the mighty rushing sound of countless strong and rapidly-plied pinions." According to Hume the breeding season for these birds in Upper India lasts from Christmas to May day. The experience of the writer is that April, May and June are the months in which to look for their nests. However, in justice to Hume, it must be said that recently Mr. A. J. Currie found a nest, containing eggs, in February.

In April the green pigeons pair and build slender cradles, high up in mango trees, in which two white eggs are laid.

The songster of the house-top—the brown rock-chat (*Cercomela fusca*)—makes sweet music throughout the month for the benefit of his spouse, who is incubating four pretty pale-blue eggs in a nest built on a ledge in an outhouse or on the sill of a clerestory window. This bird, which is thought by some to be a near relative of the sparrow of the Scriptures, is clothed in plain brown and seems to suffer from St. Vitus' dance in the tail. Doubtless it is often mistaken for a hen robin. For this mistake there is no excuse, because the rock-chat lacks the brick-red patch under the tail.

April is the month in which to look for two exquisite little nests—those of the white-eye (*Zosterops palpebrosa*) and the iora (*Aegithina tiphia*). White-eyes are minute greenish-yellow birds with a conspicuous ring of white feathers round the eye. They go about in flocks. Each individual utters unceasingly a plaintive cheeping note by means of which it keeps its fellows apprised of its whereabouts. At the breeding season, that is to say in April and May, the cock sings an exceedingly sweet, but very soft, lay of six or seven notes. The nest is a cup, about 2½ inches in diameter and ¾ of an inch in depth. It is usually suspended, like a hammock, from the fork of a branch; sometimes it is attached to the end of a single bough; it then looks like a ladle, the bough being the handle. It is composed of cobweb, roots, hair and other soft materials. Three or four tiny pale-blue eggs are laid.

The iora is a feathered exquisite, about the size of a tomtit. The cock is arrayed in green, black and gold; his mate is gowned in green and yellow.

The iora has a great variety of calls, of these a soft and rather plaintive long-drawn-out whistle is uttered most frequently in April and May.

In shape and size the nest resembles an after-dinner coffee cup. It is beautifully woven, and, like those of the white-eye and fantail flycatcher, covered with cobweb; this gives it a very neat appearance. In it are laid two or three eggs of salmon hue with reddish-brown and purple-grey blotches.

Throughout April the sprightly tailor-birds are busy with their nests. The tailor-bird (*Orthotomus sutorius*) is a wren with a long tail. In the breeding season the two median caudal feathers of the cock project as bristles beyond the others. The nest is a wonderful structure. Having selected a suitable place, which may be a bush in a garden or a pot plant in a verandah, the hen tailor-bird proceeds to make, with her sharp bill, a series of punctures along the margins of one or more leaves. The punctured edges are then drawn together, by means of strands of cobweb, to form a purse or pocket. When this has been done the frail bands of cobweb, which hold the edges of the leaves *in situ*, are strengthened by threads of cotton. Lastly, the purse is cosily lined with silk-cotton down or other soft material. Into the cradle, thus formed, three or four white eggs, speckled with red, find their way.

In April cavities in trees and buildings suitable for nesting purposes are at a premium owing to the requirements of magpie-robins, brahminy mynas, common mynas, yellow-throated sparrows and rollers. Not uncommonly three or four pairs of birds nest in one weather-beaten old tree.

Bank-mynas, white-breasted kingfishers, bee-eaters and a few belated sand-martins are nesting in sandbanks in cavities which they themselves have excavated. The nests of the kingfisher and the sand-martin have already been described, that of the bank-myna belongs to May rather than to April.

Bee-eaters working at the nest present a pleasing spectacle. The sexes excavate turn about. The site chosen may be a bunker on the golf links, the butts on the rifle range, a low mud boundary between two fields, or any kind of bank. The sharp claws of the bee-eaters enable the birds to obtain a foothold on an almost vertical surface; this foothold is strengthened by the tail which, being stiff, acts as a third leg. In a surprisingly short time a cavity large enough to conceal the bird completely is formed. The bee-eater utilises the bill as pickaxe and the feet as ejectors. The little clouds of sand that issue at short intervals from each cavity afford evidence of the efficacy of these implements and the industry of those that use them.

Two of the most charming birds in India are now occupied with family cares. These are both black-and-white birds—the magpie-robin (*Copsychus saularis*) and the pied wagtail (*Motacilla maderaspatensis*). The former has already been noticed as the best songster in the plains of India. The pattern of its plumage resembles that of the common magpie; this explains its English name. The hen is grey where the cock is black, otherwise there is no external difference between the sexes. For some weeks the cock has been singing lustily, especially in the early morning and late afternoon. In April he begins his courtship. His display is a simple affair—mere tail-play; the tail is expanded into a fan, so as to show the white outer feathers, then it is either raised and lowered alternately, or merely held depressed. Normally the tail is carried almost vertically. The nest is invariably placed in a cavity of a tree or a building.

The pied wagtail always nests near water. If not on the ground, the nursery rests on some structure built by man.

A visit to a bridge of boats in April is sure to reveal a nest of this charming bird. Hume records a case of a pair of pied wagtails nesting in a ferry-boat. This, it is true, was seldom used, but did occasionally cross the Jumna. On such occasions the hen would continue to sit, while the cock stood on the gunwale, pouring forth his sweet song, and made, from time to time, little sallies over the water after a flying gnat. Mr. A. J. Currie found at Lahore a nest of these wagtails in a ferry-boat in daily use; so that the birds must have selected the site and built the nest while the boat was passing to and fro across the river!

Yet another black-and-white bird nests in April. This is the pied bush-chat (*Pratincola caprata*). The cock is black all over, save for a white patch on the rump and a bar of white in the wing. He delights to sit on a telegraph wire or a stem of elephant grass and there make cheerful melody. The hen is a dull reddish-grey bird. The nest is usually placed in a hole in the ground or a bank or a wall, sometimes it is wedged into a tussock of grass.

Allied to the magpie-robin and the pied bush-chat is the familiar Indian robin (*Thamnobia cambayensis*), which, like its relatives, is now engaged in nesting operations. This species constructs its cup-shaped nest in all manner of strange places. Spaces in stacks of bricks, holes in the ground or in buildings, and window-sills are held in high esteem as nesting sites. The eggs are not easy to describe because they display great variation. The commonest type has a pale green shell, speckled with reddish-brown spots, which are most densely distributed at the thick end of the egg.

Many of the grey partridges (*Francolinus pondicerianus*) are now nesting. This species is somewhat erratic in respect of its breeding season. Eggs have been taken in February, March, April, May, June, September, October, and November. The April eggs, however, outnumber those of all the other months put together. The nest is a shallow depression in the ground, lined with grass, usually under a bush. From six to nine cream-coloured eggs are laid.

Another bird which is now incubating eggs on the ground is the did-he-do-it or red-wattled lapwing (*Sarcogrammus indicus*). The curious call, from which this plover derives its popular name, is familiar to every resident in India. This species nests between March and August. The 122 eggs in the possession of Hume were taken, 12 in March, 46 in April, 24 in May, 26 in June, 4 in July, and 8 in August. Generally in a slight depression on the ground, occasionally on the ballast of a rail-road, four pegtop-shaped eggs are laid; these are, invariably, placed in the form of a cross, so that they touch each other at their thin ends. They are coloured like those of the common plover. The yellow-wattled lapwing (*Sarciophorus malabaricus*), which resembles its cousin in manners and appearance, nests in April, May and June.

The nesting season of the various species of sand-grouse that breed in India is now beginning. These birds, like lapwings, lay their eggs on the ground.

In April one may come across an occasional nest of the pied starling, the king-crow, the paradise flycatcher, the grey hornbill, and the oriole, but these are exceptions. The birds in question do not as a rule begin to nest until May, and their doings accordingly are chronicled in the calendar for that month.

MAY

> The melancholy days are come, the saddest of the year.
>
> *The Minstrelsy of the Woods.*

> Low from the brink the waters shrink;
> The deer all snuff for rain;
> The panting cattle search for drink
> Cracked glebe and dusty plain;
> The whirlwind, like a furnace blast,
> Sweeps clouds of darkening sand.
>
> WATERFIELD. *Indian Ballads.*

> Now the burning summer sun
> Hath unchalleng'd empire won
> And the scorching winds blow free,
> Blighting every herb and tree.
>
> R. T. H. GRIFFITH.

May in the plains of India! What unpleasant memories it recalls! Stifling nights in which sleep comes with halting steps and departs leaving us unrefreshed. Long, dreary days beneath the punkah in a closed bungalow which has ceased to be enlivened by the voices of the children and the patter of their little feet. Hot drives to office, under a brazen sky from which the sun shines with pitiless power, in the teeth of winds that scorch the face and fill the eyes with dust.

It is in this month of May that the European condemned to existence in the plains echoes the cry of the psalmist: "Oh that I had wings like a dove! for then would I fly away, and be at rest"— in the Himalayas. There would I lie beneath the deodars and, soothed by the rustle of their wind-caressed branches, drink in the pure cool air and listen to the cheerful double note of the cuckoo. The country-side in the plains presents a sorry spectacle. The gardens that had some beauty in the cold weather now display the abomination of desolation—a waste of shrivelled flowers, killed by the relentless sun. The spring crops have all been cut and the whole earth is dusty brown save for a few patches of young sugar-cane and the dust-covered verdure of the mango topes. It is true that the gold-mohur trees and the Indian laburnums are in full flower and the air is heavily laden with the strong scent of the *nim* blossoms, but the heat is so intense that the European is able to enjoy these gifts of nature only at dawn. Nor has the ripening jack-fruit any attractions for him. He is repelled by its overpowering scent and sickly flavour. Fortunately the tastes of all men are not alike. In the eyes of the Indian this fruit is a dish fit to be set before the gods. The *pipal* trees, which are covered with tender young leaves, now offer to the birds a feast in the form of numbers of figs, no larger than cranberries. This generous offer is greedily accepted by green pigeons, mynas and many other birds which partake with right goodwill and make much noise between the courses. No matter how intense the heat be, the patient cultivator issues forth with his cattle before sunrise and works at his threshing floor until ten o'clock, then he seeks the comparative coolness of the mango tope and sleeps until the sun is well on its way to the western horizon, when he resumes the threshing of the corn, not ceasing until the shades of night begin to steal over the land.

The birds do not object to the heat. They revel in it. It is true that in the middle of the day even they seek some shady tree in which to enjoy a siesta and await the abatement of the heat of the blast furnace in which they live, move and have their being. The long day, which begins for them before 4 a.m., rather than the intense heat, appears to be the cause of this midday sleep. Except during this period of rest at noon the birds are more lively than they were in April.

The breeding season is now at its height. In May over five hundred species of birds nest in India. No individual is likely to come across all these different kinds of nests, because, in order to do so, that person would have to traverse India from Peshawar to Tinnevelly and from Quetta to Tenasserim. Nevertheless, the man who remains in one station, if he choose to put forth a little energy and defy the sun, may reasonably expect to find the nests of more than fifty kinds of birds. Whether he be energetic or the reverse he cannot fail to hear a great many avian sounds both by day and by night. In May the birds are more vociferous than at any other time of year. The fluty cries of the koel and the vigorous screams of the brain-fever bird penetrate the closed doors of the bungalow, as do, to a less extent, the chatter of the seven sisters, the calls of the mynas, the *towee, towee, towee* of the tailor-bird, the *whoot, whoot, whoot* of the crow-pheasant, the monotonous notes of the coppersmith and the green barbet, the *uk, uk, uk* of the hoopoe, the cheerful music of the fantail flycatcher, the three sweet syllables of the iora—*so be ye*, the *tee, tee, tee, tee* of the nuthatch, the liquid whistle of the oriole and, last but not least, the melody of the magpie-robin. The calls of the hoopoe and nuthatch become less frequent as the month draws to a close; on the other hand, the melody of the oriole gains in strength.

As likely as not a pair of blue jays has elected to rear a brood of young hopefuls in the chimney or in a hole in the roof. When this happens the human occupant of the bungalow is apt to be driven nearly to distraction by the cries of the young birds, which resemble those of some creature in distress, and are uttered with "damnable reiteration."

All these sounds, however, reach in muffled form the ear of a human being shut up in a bungalow; hence it is the voices of the night rather than those of the day with which May in India is associated. Most people sleep out of doors at this season, and, as the excessive heat makes them restless, they have ample opportunity of listening to the nightly concert of the feathered folk. The most notable performers are the cuckoos. These birds are fully as nocturnal as the owls. The brain-fever bird (*Hierococcyx varius*) is now in full voice, and may be heard, both by day and by night, in all parts of Northern India, east of Umballa. This creature has two calls. One is the eternal "brain-fever, *brain-fever*, BRAIN-FEVER," each "brain-fever" being louder and pitched in a higher key than the previous one, until the bird reaches its top note. The other call consists of a volley of descending notes, uttered as if the bird were unwinding its voice after the screams of "brain-fever." The next cuckoo is not one whit less vociferous than the last. It is known as the Indian koel (*Eudynamis honorata*). This noble fowl has three calls, and it would puzzle anyone to say which is the most powerful. The usual cry is a crescendo *ku-il, ku-il, ku-il,* which to Indian ears is very sweet-sounding. Most Europeans are agreed that it is a sound of which one can have too much. The second note is a mighty avalanche of yells and screams, which Cunningham has syllabised as *Kúk, kuu, kuu, kuu, kuu, kuu.* The third cry, which is uttered only occasionally, is a number of shrill shrieks: *Hekaree, karee, karee, karee.*

The voice of the koel is heard throughout the hours of light and darkness in May, so that one wonders whether this bird ever sleeps. The second call is usually reserved for dawn, when the bird is most vociferous. This cry is particularly exasperating to Europeans, since it often awakens them rudely from the only refreshing sleep they have enjoyed, namely, that obtained at the time when the temperature is comparatively low. The koel extends into the Punjab and is heard throughout Northern India.

The third of the cuckoos which enlivens the hot weather in the plains is the Indian cuckoo (*Cuculus micropterus*). This species dwells chiefly in the Himalayas, but late in April or early in May certain individuals seek the hot plains and remain there for some months. They do not extend very far into the peninsula, being numerous only in the sub-Himalayan tracts as far south as Fyzabad. The call of this cuckoo is melodious and easily recognised. Indians represent it as *Bouto-taku,* while some Englishmen maintain that the bird says "I've lost my love." To the writer's mind the cry is best represented by the words *wherefore, wherefore,* repeated with musical cadence. This bird does not usually call much during the day. It uplifts its voice about two hours before sunset and continues calling intermittently until some time after sunrise. The note is often uttered while the bird is on the wing.

Scarcely less vociferous than the cuckoos are the owls. Needless to state that the tiny spotted owlets make a great noise in May. They are loquacious throughout the year, especially on moonlight nights. Nor do they wait for the setting of the sun until they commence to pour forth what Eha terms a "torrent of squeak and chatter and gibberish."

Almost as abundant as the spotted owlet is the jungle owlet (*Glaucidium radiatum*). This species, like the last-mentioned, does not confine its vocal efforts to the hot weather. It is vociferous throughout the year; however, special mention must be made of it in connection with the month of May, because it is not until a human being sleeps out of doors that he takes much notice of the bird.

The note of this owl is very striking. It may be likened to the noise made by a motor cycle when it is being started. It consists of a series of dissyllables, low at first with a pause after each, but gradually growing in intensity and succeeding one another at shorter intervals, until the bird seems to have got fairly into its stride, when it pulls up with dramatic suddenness. Tickell thus syllabises its call: *Turtuck, turtuck, turtuck, turtuck, turtuck, tukatu, chatatuck, atuckatuck.*

Another sound familiar to those who sleep out of doors at this season is a low, soft "what," repeated at intervals of about a minute.

The writer ascribes this call to the collared scops owl (*Scops bakkamoena*). Mr. A. J. Currie, however, asserts that the note in question is that emitted by spotted owlets (*Athene brama*) when they have young. He states that he has been quite close to the bird when it was calling.

A little patient observation will suffice to decide the point at issue.

It is easy to distinguish between the two owls, as the scops has aigrettes or "horns," which the spotted owlet lacks.

The nightjars help to swell the nocturnal chorus. There are seven or eight different species in India, but of these only three are commonly heard and two of them occur mainly in forest tracts. The call of the most widely-distributed of the Indian goatsuckers—*Caprimulgus asiaticus*, the common Indian nightjar—is like unto the sound made by a stone skimming over ice. Horsfield's goatsucker is a very vociferous bird. From March till June it is heard wherever there are forests. As soon as the shadows of the evening begin to steal across the sky its loud *chuk, chuk, chuk, chuk, chuk* cleaves the air for minutes together. This call to some extent replaces by night the *tonk, tonk, tonk* of the coppersmith, which is uttered so persistently in the day-time. In addition to this note Horsfield's nightjar emits a low soft *chur, chur, chur*.

The third nightjar, which also is confined chiefly to forest tracts, is known as Franklin's nightjar (*C. monticolus*). This utters a harsh *tweet* which at a distance might pass for the chirp of a canary with a sore throat.

Other sounds heard at night-time are the plaintive *did-he-do-it pity-to-do-it* of the red-wattled lapwing (*Sarcogrammus indicus*), and the shrill calls of other plovers.

As has already been said, the nesting season is at its height in May. With the exception of the paroquets, spotted owlets, nuthatches, black vultures and pied kingfishers, which have completed nesting operations for the year, and the golden-backed woodpeckers and the cliff-swallows, which have reared up their first broods, the great majority of the birds mentioned as having nests or young in March or April are still busily occupied with domestic cares.

May marks the close of the usual breeding season for the jungle crows, skylarks, crested larks, finch-larks, wood-shrikes, yellow-throated sparrows, sand-martins, pied wagtails, green barbets, coppersmiths, rollers, green bee-eaters, white-breasted kingfishers, scavenger vultures, tawny eagles, kites, shikras, spur-winged plovers, little ringed plovers, pied woodpeckers, night herons and pied chats. In the case of the tree-pies, cuckoo-shrikes, seven sisters, bank-mynas and blue-tailed bee-eaters the nesting season is now at its height. All the following birds are likely to have either eggs or nestlings in May: the white-eyes, ioras, bulbuls, tailor-birds, shrikes, brown rock-chats, Indian robins, magpie-robins, sunbirds, swifts, nightjars, white-eyed buzzards, hoopoes, green pigeons, blue rock-pigeons, doves, sparrows, the red and yellow wattled lapwings, minivets, wire-tailed swallows, red-headed merlins, fantail flycatchers, pipits, sand-grouse and grey partridges. The nests of most of these have been described already.

In the present month several species begin nesting operations. First and foremost among these is the king-crow or black drongo (*Dicrurus ater*). No bird, not even the roller, makes so much ado about courtship and nesting as does the king-crow, of which the love-making was described last month. A pair of king-crows regards as its castle the tree in which it has elected to construct a nest. Round this tree it establishes a sphere of influence into which none but a favoured few birds may come. All intruders are forthwith set upon by the pair of little furies, and no sight is commoner at this season than that of a crow, a kite, or a hawk being chased by two irate drongos. The nest of the king-crow is a small cup, wedged into the fork of a branch high up in a tree.

The Indian oriole (*Oriolus kundoo*) is one of the privileged creatures allowed to enter the dicrurian sphere of influence, and it takes full advantage of this privilege by placing its nest almost invariably in the same tree as that of the king-crow. The oriole is a timid bird and is glad to rear up its family under the ægis of so doughty a warrior as the Black Prince of the Birds. The nest of the oriole is a wonderful structure. Having selected a fork in a suitable branch, the nesting bird tears off a long strip of soft pliable bark, usually that of the mulberry tree. It proceeds to wind one end of this strip round a limb of the forked branch, then the other end is similarly bound to the other limb. A second and a third strip of bark are thus dealt with, and in this manner a cradle or hammock is formed. On it a slender cup-shaped nest is superimposed. This is composed of grasses and fibres, some of which are wound round the limbs of the forked branch, while others are made fast to the strands of bark. The completed nest is nearly five inches in diameter. From below it looks like a ball of dried grass wedged into the forked branch.

The oriole lays from two to four white eggs spotted with dull red. The spots can be washed off by water; sometimes their colour "runs" while they are in the nest, thereby imparting a pink hue to the whole shell. Both sexes take part in nest construction, but the hen alone appears to incubate. She is a very shy creature, and is rarely discovered actually sitting, because she leaves the nest with a little cry of alarm at the first sound of a human footfall.

May and June are the months in which to look for the nests of that superb bird—the paradise flycatcher (*Terpsiphone paradisi*). This is known as the rocket-bird or ribbon-bird because of the two long fluttering tail feathers possessed by the cock. The hen has the appearance of a kind of bulbul, being chestnut-hued with a white breast and a metallic blue-black crest. For the first year of their existence the young cocks resemble the hens in appearance. Then the long tail feathers appear. In his third year the cock turns white save for the black-crested head. This species spends the winter in South India. In April it migrates northwards to summer in the shady parts of the plains of Bengal, the United Provinces and the Punjab, and on the lower slopes of the Himalayas. The nest is a deep, untidy-looking cup, having the shape of an inverted cone. It is always completely covered with cocoons and cobweb. It is usually attached to one or more of the lower branches of a tree. Both sexes work at the nest and take part in incubation. The long tail feathers of the sitting cock hang down from the nest like red or white satin streamers according to the phase of his plumage. In the breeding season the cock sings a sweet little lay—an abridged version of that of the fantail flycatcher. When alarmed both the cock and the hen utter a sharp *tschit*.

May is perhaps the proper month in which to describe the nesting of the various species of myna.

According to Hume the normal breeding season of the common myna (*Acridotheres tristis*) lasts from June to August, during which period two broods are reared. This is not correct. The nesting season of this species begins long before June. The writer has repeatedly seen mynas carrying twigs and feathers in March, and has come across nests containing eggs or young birds in both April and May. June perhaps is the month in which the largest numbers of nests are seen. The cradle of the common myna is devoid of architectural merit. It is a mere conglomeration of twigs, grass, rags, bits of paper and other oddments. The nesting material is dropped haphazard into a hole in a tree or building, or even on to a ledge in a verandah. Four beautiful blue eggs are laid.

At Peshawar Mr. A. J. Currie once found four myna's eggs in a deserted crows' nest in a tree.

As has already been stated, the nest of the bank-myna (*A. ginginianus*) is built in a hole in a well, a sandbank, or a cliff. The birds breed in colonies; each pair excavates its own nest by means of beak and claw. Into the holes dug out in this manner the miscellaneous nesting materials are dropped pell-mell after the manner of all mynas. The breeding season of this species lasts from April to July, May being the month in which most eggs are laid.

The black-headed or brahminy myna (*Temenuchus pagodarum*) usually begins nesting operations about a month later than the bank-myna; its eggs are most often taken in June. The nest, which is an untidy, odoriferous collection of rubbish, is always in a cavity. In Northern India a hole in a tree is usually selected; in the South buildings are largely patronised. Some years ago the writer observed a pair of these birds building a nest in a hole made in the masonry for the passage of the lightning conductor of the Church in Fort St. George, Madras.

May marks the commencement of the breeding season of the pied starlings (*Sturnopastor contra*). In this month they begin to give vent with vigour to their cheerful call, which is so pleasing as almost to merit the name of song.

Throughout the rains they continue to make a joyful noise. Not that they are silent at other seasons; they call throughout the year, but, except at the breeding period, their voices are comparatively subdued.

The nest is a bulky, untidy mass of straw, roots, twigs, rags, feathers and such-like things. It is placed fairly low down in a tree.

Many of these nests are to be seen in May, but the breeding season is at its height in June and July.

The grey hornbills (*Lophoceros birostris*) are now seeking out holes in which to deposit their eggs. The hen, after having laid the first egg, does not emerge from the nest till the young are ready to fly. During the whole of this period she is kept a close prisoner, the aperture to the nest cavity having been closed by her mate and herself with their own droppings, a small chink alone being left through which she is able to insert her beak in order to receive the food brought to her by the cock.

Mr. A. J. Currie gives an interesting account of a grey hornbill's nest he discovered at Lahore in 1910. About the middle of April he noticed a pair of paroquets nesting in a hole in a tree. On April 28th he saw a hornbill inspecting the hole, regardless of the noisy protests of the paroquets. On the 30th he observed that the hole had become smaller, and suspected that the hornbills had taken possession. On May 1st all that was left of the hole was a slit. On May 6th Mr. Currie watched the cock hornbill feeding the hen.

First the male bird came carrying a fig in his bill. Seeing human beings near the nest, he did not give the fig to the hen but swallowed it and flew off. Presently the cock reappeared with a fig which he put into the slit in the plastering; after he had parted with the fig he began to feed the hen by bringing up food from his crop. During the process the beak of the hen did not appear at the slit.

On May 7th Mr. Currie opened out the nest. The hole was sixteen feet from the ground and the orifice had a diameter of three inches; all of this except a slit, broadest at the lower part, was filled up by plaster. This plaster was odourless and contained embedded in it a number of fig seeds.

The nest hole was capacious, its dimensions being roughly 1 foot by 1 foot by 2 feet. From the bottom five handfuls of pieces of dry bark were extracted. Three white eggs were found lying on these pieces of bark. The sitting hen resented the "nest-breaking," and, having pecked viciously at the intruder, tried to escape by climbing up to the top of the nest hole. She was dragged out of her retreat by the beak, after an attempt to pull her out by the tail had resulted in all her tail feathers coming away in her captor's hand!

The young green parrots have all left their nests and are flying about in noisy flocks. They may be distinguished from the adults by the short tail and comparatively soft call.

Most pairs of hoopoes are now accompanied by at least one young bird which is almost indistinguishable from the adults. The young birds receive, with squeaks of delight, the grubs or caterpillars proffered by the parents. Occasionally a pair of hoopoes may be seen going through the antics of courtship preparatory to raising a second brood.

In scrub-jungle parties of partridges, consisting of father, mother and five or six little chicks, wander about.

As the shades of night begin to fall family parties of spotted owlets issue from holes in trees or buildings. The baby birds squat on the ground in silence, while the parents make sallies into the air after flying insects which they bring to the young birds.

The peafowl and sarus cranes are indulging in the pleasures of courtship. The young cranes, that were hatched out in the monsoon of last year, are now nearly as big as their parents, and are well able to look after themselves; ere long they will be driven away and made to do so. The display of the sarus is not an elaborate process. The cock turns his back on the hen and then partially opens his wings, so that the blackish primaries droop and the grey secondary feathers are arched. In this attitude he trumpets softly.

The water-hens have already begun their uproarious courtship. Their weird calls must be heard to be appreciated. They consist of series of *kok*, *koks* followed by roars, hiccups, cackles and gurgles.

Black partridges, likewise, are very noisy throughout the month of May. Their nesting season is fast approaching.

Even as April showers in England bring forth May flowers, so does the April sunshine in India draw forth the marriage adornments of the birds that breed in the rains. The pheasant-tailed jacanas are acquiring the long tail feathers that form the wedding ornaments of both sexes.

The various species of egret and the paddy bird all assume their nuptial plumes in May.

In the case of the egret these plumes are in great demand and are known to the plumage trade as "ospreys."

The plumes in question consist of long filamentous feathers that grow from the neck of the egret and also from its breast. In most countries those who obtain these plumes wait until the birds are actually nesting before attempting to secure them, taking advantage of the fact that egrets nest in colonies and of the parental affection of the breeding birds. A few men armed with guns are able to shoot every adult member of the colony, because the egrets continue to feed their young until they are shot. As the plumes of these birds are worth nearly their weight in gold, egrets have become extinct in some parts of the world.

The export of plumage from India is unlawful, but this fact does not prevent a very large feather trade being carried on, since it is not difficult to smuggle "ospreys" out of the country.

Doubtless the existing Notification of the Government of India, prohibiting the export of plumage, has the effect of checking, to some extent, the destruction of egrets, but there is no denying the fact that many of the larger species are still shot for their plumes while breeding.

In the case of cattle-egrets (*Bubulcus coromandus*) the custom of shooting them when on the nest has given place to a more humane and more sensible method of obtaining their nuchal plumes. These, as we have seen, arise early in May, but the birds do not begin to nest until the end of June. The cattle-egret is gregarious; it is the large white bird that accompanies cattle in order to secure the insects put up by the grazing quadrupeds. Taking advantage of the social habits of these egrets the plume-hunters issue forth early in May and betake themselves, in parties of five or six, to the villages where the birds roost. Their apparatus consists of two nets, each some eight feet long and three broad. These are laid flat on the ground in shallow water, parallel to one another, about a yard apart. The inner side of each net is securely pegged to the ground. By an ingenious arrangement of sticks and ropes a man, taking cover at a distance of twenty or thirty yards, by giving a sharp pull at a pliable cane, can cause the outer parts of each net to spring up and meet to form an enclosure which is, in shape, not unlike a sleeping-pal tent. When the nets have been set in a pond near the trees where the cattle-egrets roost at night and rest in the day-time, two or three decoy birds—captured egrets with their eyes sewn up to prevent them struggling or trying to fly away—are tethered in the space between the two nets; these last, being laid flat under muddy water, are invisible. Sooner or later an egret in one of the trees near by, seeing some of its kind standing peacefully in the water, alights near them. Almost before it has touched the ground the cane is pulled and the egret finds itself a prisoner. One of the bird-catchers immediately runs to the net, secures the victim, opens out its wings, and, holding each of these between the big and the second toe, pulls out the nuchal plumes. This operation lasts about five seconds. The bird is then set at liberty, far more astonished than hurt. It betakes itself to its wild companions, and the net is again set. Presently another egret is caught and divested of its plumes, and the process continues all day.

The bird-catchers spend six weeks every year in obtaining cattle-egret plumes in this manner. They sell the plumes to middle-men, who dispose of them to those who smuggle them out of India.

If stuffed birds were used as decoys and the plumes of the captured birds were snipped off with scissors instead of being pulled out, the operation could be carried on without any cruelty, and, if legalised and supervised by the Government, it could be made a source of considerable revenue.

JUNE

'Tis raging noon; and, vertical, the sun
Darts on the head direct his forceful rays;
O'er heaven and earth, far as the ranging eye
Can sweep, a dazzling deluge reigns; and all
From pole to pole is undistinguish'd blaze.

All-conquering heat, oh, intermit thy wrath,
And on my throbbing temples potent thus
Beam not so fierce! incessant still you flow,
And still another fervent flood succeeds.
Pour'd on the head profuse. In vain I sigh,

Thrice happy he who on the sunless side
Of a romantic mountain, forest crown'd
Beneath the whole collected shade reclines.

J. THOMSON.

With dancing feet glad peafowl greet
Bright flash and rumbling cloud;
Down channels steep red torrents sweep;
The frogs give welcome loud;

No stars in skies, but lantern-flies
Seem stars that float to earth.

WATERFIELD. *Indian Ballads.*

There are two Indian Junes—the June of fiction and the June of fact. The June of fiction is divided into two equal parts—the dry half and the wet half. The former is made up of hot days, dull with dust haze, when the shade temperature may reach 118°, and of oppressive nights when the air is still and stagnant and the mercury in the thermometer rarely falls below 84°. Each succeeding period of four-and-twenty hours seems more disagreeable and unbearable than its predecessor, until the climax is reached about the 15th June, when large black clouds appear on the horizon and roll slowly onwards, accompanied by vivid lightning, loud peals of thunder and torrential rain. In the June of fact practically the whole month is composed of hot, dry, dusty, oppressive days; for the monsoon rarely reaches Northern India before the last week of the month and often tarries till the middle of July, or even later.

The first rain causes the temperature to fall immediately. It is no uncommon thing for the mercury in the thermometer to sink 20 degrees in a few minutes. While the rain is actually descending the weather feels refreshingly cool in contrast to the previous furnace-like heat. Small wonder then that the advent of the creative monsoon is more heartily welcomed in India than is spring in England. No sound is more pleasing to the human ear than the drumming of the first monsoon rain.

But alas! the physical relief brought by the monsoon is only temporary. The temperature rises the moment the rain ceases to fall, and the prolonged breaks in the rains that occur every year render the last state of the climate worse than the first. The air is so charged with moisture that it cannot absorb the perspiration that emanates from the bodies of the human beings condemned to existence in this humid Inferno. For weeks together we live in a vapour-bath, and to the physical discomfort of perpetual clamminess is added the irritation of prickly heat.

Moreover, the rain brings with it myriads of torments in the form of termites, beetles, stinking bugs, flies, mosquitoes and other creeping and flying things, which bite and tease and find their way into every article of food and drink. The rain also awakens from their slumbers the frogs that have hibernated and æstivated in the sun-baked beds of dried-up ditches and tanks. These awakened amphibia fill the welkin with their croakings, which take the place of the avian chorus at night. The latter ceases with dramatic abruptness with the first fall of monsoon rain. During the monsoon the silence of the night is broken only by the sound of falling raindrops, or the croaking of the frogs, the stridulation of crickets innumerable, and the owlet's feeble call. Before the coming of the monsoon the diurnal chorus of the day birds begins to flag because the nesting season for many species is drawing to a close. The magpie-robin still pours forth his splendid song, but the quality of the music in the case of many individuals is already beginning to fall off. The rollers, which are feeding their young, are far less noisy than they were at the time of courtship. The barbets and coppersmiths, although not so vociferous as formerly, cannot, even in the monsoon, be charged with hiding their lights under a bushel. Towards the end of June the *chuk, chuk, chuk, chuk, chuk* of Horsfield's nightjar is not often heard, but the bird continues to utter its soft churring note. The iora's cheerful calls still resound through the shady mango tope. The sunbirds, the fantail flycatchers, the orioles, the golden-backed woodpeckers, the white-breasted kingfishers and the black partridges call as lustily as ever, and the bulbuls continue to twitter to one another "stick to it!" With the first fall of rain the tunes of the paradise flycatchers and the king-crows change. The former now cry "Witty-ready wit," softly and gently, while the calls of the latter suddenly become sweet and mellow.

Speaking generally, the monsoon seems to exercise a sobering, a softening influence on the voices of the birds. The pied myna forms the one exception; he does not come into his full voice until the rains have set in.

The monsoon transfigures the earth. The brown, dry, hard countryside, with its dust-covered trees, becomes for the time being a shallow lake in which are studded emerald islets innumerable. Stimulated by the rain many trees put forth fresh crops of leaves. At the first break in the downpour the cultivators rush forth with their ploughs and oxen to prepare the soil for the autumn crops with all the speed they may.

There is much to interest the ornithologist in June.

Of the birds whose nests have been previously described the following are likely to have eggs or young: white-eyes, ioras, tailor-birds, king-crows, robins, sparrows, tree-pies, seven sisters, cuckoo-shrikes, Indian wren-warblers (second brood), sunbirds (second brood), swifts, fantail flycatchers (second brood), orioles, paradise flycatchers, grey horn-bills, and the various mynas, bulbuls, butcher-birds, doves, pigeons and lapwings. The following species have young which either are in the nest or have only recently left it: roller, hoopoe, brown rock-chat, magpie-robin, coppersmith, green barbet, nightjar, white-eyed buzzard, pipit, wire-tailed swallow, white-breasted kingfisher, grey partridge, kite, golden-backed woodpecker (second brood), and the several species of bee-eater and lark.

With June the breeding season for the blue rock and green pigeons ends. In the *sal* forests the young jungle-fowl have now mostly hatched out and are following the old hens, or feeding independently.

Some of the minivets are beginning to busy themselves with a second brood.

The breeding operations of a few species begin in June.

Chief of these is that arch-villain *Corvus splendens*—the Indian house-crow. Crows have no fine feathers, hence the cocks do not "display" before the hens. To sing they know not how. Their courtship, therefore, provides a feast for neither the eye nor the ear of man. The lack of ornaments and voice perhaps explains the fact that among crows there is no noisy love-making. Crows make a virtue of necessity. Any attempt at courtship after the style of the costermonger is resented by the whole corvine community. The only amorous display permitted in public is head-tickling. The cock and the hen perch side by side, one ruffles the feathers of the neck, the other inserts its bill between the ruffled feathers of its companion and gently tickles its neck, to the accompaniment of soft gurgles.

Crows are the most intelligent of birds. Like the other fowls of the air in which the brain is well developed, they build rough untidy nests—mere platforms placed in the fork of a branch of almost any kind of tree. The usual materials used in nest-construction are twigs, but crows do not limit themselves to these. They seem to take a positive pride in pressing into service materials of an uncommon nature. Cases are on record of nests composed entirely of spectacle-frames, wires used for the fixing of the corks of soda-water bottles, or pieces of tin discarded by tinsmiths.

Four, five or six eggs are laid; these are of a pale greenish-blue hue, speckled or flaked with sepia markings. The hen alone collects the materials for the nest, but the cock supervises her closely, following her about and criticising her proceedings as she picks up twigs and works them into the nest.

From the time of the laying of the first egg until the moment of the departure of the last young bird, one or other of the parents always mounts guard over the nest, except when they are chasing a koel. Crows are confirmed egg-lifters and chicken-stealers; they apply their standard of morality to other birds, and, in consequence, never leave their own offspring unguarded. A crow's nest at which there is no adult crow certainly contains neither eggs nor young birds.

As has already been stated, crows spend, much time in teasing and annoying other birds. Retribution overtakes them in the nesting season. The Indian koel (*Eudynamis honorata*) cuckolds them. The crows either are aware of this or have an instinctive dislike to this cuckoo. The sight of the koel affects a crow in much the same way as a red cloth irritates a bull. One of these cuckoos has but to perch in a tree that contains a crow's nest and begin calling in order to make both the owners of the nest attack him. The koel takes full advantage of this fact. The cock approaches the nest and begins uttering his fluty *kuil, kuil*. The crows forthwith dash savagely at him. He flies off pursued by them. He can easily outdistance his pursuers, but is content to keep a lead of a few feet, crying *pip-pip* or *kuil-kuil*, and thus he lures the parent crows to some distance. No sooner are their backs turned than the hen koel slips quietly into the nest and deposits an egg in it. If she have time she carries off or throws out one or more of the legitimate eggs. When the crows return to the nest, having failed to catch the cock koel, they do not appear to notice the trick played upon them, although the koel's egg is smaller than theirs and of an olive-green colour. Through the greater part of June and July the koels keep the crows busy chasing them. Something approaching pandemonium reigns in the neighbourhood of a colony of nesting crows: from dawn till nightfall the shrieks and yells of the koels mingle with the harsh notes of the crows.

Sometimes the crows return from the chase of the cock koel before the hen is ready, and surprise her in the nest; then they attack her. She flees in terror, and is followed by the corvi. Her screams when being thus pursued are loud enough to awaken the Seven Sleepers. She has cause for alarm, for, if the raging crows catch her, they will assuredly kill her. Such a tragedy does sometimes occur.

Not infrequently it happens that more than one koel's egg is laid in a crow's nest.

The incubation period of the egg of the koel is shorter than that of the crow, the consequence is that when, as usually happens, there is one of the former and several of the latter in a nest, the young koel is invariably the first to emerge. It does not attempt to eject from the nest either the legitimate eggs or the young crows when they appear on the scene. Indeed, it lives on excellent terms with its foster brethren. But to say this is to anticipate, for as a rule, neither young koels nor baby crows hatch out until July.

The crow-pheasants (*Centropus sinensis*), which are cuckoos that do not lead a parasitic existence, are now busy with nursery duties. The nest of the crow-pheasant or coucal is a massive structure, globular in shape, with the entrance at one side. Large as the nest is, it is not often discovered by the naturalist because it is almost invariably situated in the midst of an impenetrable thicket. Three or four pure-white eggs are laid.

The white-necked storks or beef-steak birds (*Dissura episcopus*) are busy at their nests in June. These birds build in large trees, usually at a distance from water. The nest is rudely constructed of twigs. It is about one and a half feet in diameter. The eggs are placed in a depression lined with straw, grass or feathers. White-necked storks often begin nest-building about the middle of May, but eggs are rarely laid earlier than the second week of June. House-crows nest at the same time of year, and they often worry the storks considerably by their impudent attempts to commit larceny of building material.

The breeding season of the paddy-birds has now fairly begun. These birds, usually so solitary in habit, often nest in small colonies, sometimes in company with night-herons. The nest is a slender platform of sticks placed high up in a tree, often in the vicinity of human habitations. Nesting paddy-birds, or pond-herons as they are frequently called, utter all manner of weird calls, the one most frequently heard being a curious gurgle.

Some of the amadavats build nests in June, but the great majority breed during the winter months.

As soon as the first rains have fallen a few of the pheasant-tailed jacanas begin nesting operations, but the greater number breed in August; for this reason their nests are described in the calendar for that month.

In June a very striking bird makes its appearance in Northern India. This is the pied crested cuckoo (*Coccystes jacobinus*). Its under parts are white, as is a bar in the wing. The remainder of the plumage is glossy black. The head is adorned by an elegant crest. The pied cuckoo has a peculiar metallic call, which is as easy to recognise as it is difficult to describe. The bird victimises, not crows, but babblers; nevertheless the corvi seem to dislike it as intensely as they dislike koels.

By the beginning of the month the great majority of the cock *bayas* or weaver-birds have assumed their black-and-golden wedding garment; nevertheless they do not as a rule begin to nest before July.

The curious excrescence on the bill of the drake *nukta* or comb-duck is now much enlarged. This betokens the approach of the nesting season for that species.

If the monsoon happen to burst early many of the birds which breed in the rains begin building their nests towards the end of June, but, in nine years out of ten, July marks the beginning of the breeding period of aquatic birds, therefore the account of their nests properly finds place in the calendar of that month, or of August, when the season is at its height.

JULY

> Alas! creative nature calls to light
> Myriads of winged forms in sportive flight,
> When gathered clouds with ceaseless fury pour
> A constant deluge in the rushing shower.
>
> Calcutta: A Poem.

In July India becomes a theatre in which Nature stages a mighty transformation scene. The prospect changes with kaleidoscopic rapidity. The green water-logged earth is for a time overhung by dull leaden clouds; this sombre picture melts away into one, even more dismal, in which the rain pours down in torrents, enveloping everything in mist and moisture. Suddenly the sun blazes forth with indescribable brilliance and shines through an atmosphere, clear as crystal, from which every particle of dust has been washed away. Fleecy clouds sail majestically across the vaulted firmament. Then follows a gorgeous sunset in which changing colours run riot through sky and clouds—pearly grey, jet black, dark dun, pale lavender, deep mauve, rich carmine, and brightest gold. These colours fade away into the darkness of the night; the stars then peep forth and twinkle brightly. At the approach of "rosy-fingered" dawn their lights go out, one by one. Then blue tints appear in the firmament which deepen into azure. The glory of the ultramarine sky does not remain long without alloy: clouds soon appear. So the scene ever changes, hour by hour and day by day. Had the human being who passes July in the plains but one window to the soul and that the eye, the month would be one of pure joy, a month spent in the contemplation of splendid dawns, brilliant days, the rich green mantle of the earth, the majesty of approaching thunderclouds, and superb sunsets. But, alas, July is not a month of unalloyed pleasure. The temperature is tolerably low while the rain is actually falling; but the moment this ceases the European is subjected to the acute physical discomforts engendered by the hot, steamy, oppressive atmosphere, the ferocity of the sun's rays, and the teasing of thousands of biting and buzzing insects which the monsoon calls into being. Termites, crickets, red-bugs, stink-bugs, horseflies, mosquitoes, beetles and diptera of all shapes and sizes arise in millions as if spontaneously generated. Many of these are creatures of the night. Although born in darkness all seem to strive after light. Myriads of them collect round every burning lamp in the open air, to the great annoyance of the human being who attempts to read out of doors after dark. The spotted owlets, the toads and the lizards, however, take a different view of the invasion and partake eagerly of the rich feast provided for them. Notwithstanding the existence of *chiks*, or gauze doors, the hexapods crowd into the lighted bungalow, where every illumination soon becomes the centre of a collection of the bodies of the insects that have been burned by the flame, or scorched by the lamp chimney. Well is it for the rest of creation that most of these insects are short-lived. The span of life of many is but a day: were it much longer human beings could hardly manage to exist during the rains. Equally unbearable would life be were all the species of monsoon insects to come into being simultaneously. Fortunately they appear in relays. Every day some new forms enter on the stage of life and several make their exit. The pageant of insect life, then, is an ever-changing one. To-day one species predominates, to-morrow another, and the day after a third. Unpleasant and irritating though these insect hosts be to human beings, some pleasure is to be derived from watching them. Especially is this the case when the termites or white-ants swarm. In the damp parts of Lower Bengal these creatures may emerge at any time of the year. In Calcutta they swarm either towards the close of the rainy season or in spring after an exceptionally heavy thunderstorm. In Madras they emerge from their hiding-places in October with the northeast monsoon. In the United Provinces the winged termites appear after the first fall of the monsoon rain in June or July as the case may be. These succulent creatures provide a feast for the birds which is only equalled by that furnished by a flight of locusts. In the case of the termites it is not only the birds that partake. The ever-vigilant crows are of course the first to notice a swarm of termites, and they lose no time in setting to work. The kites are not far behind them. These great birds sail on the outskirts of the flight, seizing individuals with their claws and transferring them to the beak while on the wing. A few king-crows and bee-eaters join them. On the ground below magpie-robins, babblers, toads, lizards, musk-rats and other terrestrial creatures make merry. If the swarm comes out at dusk, as often happens, bats and spotted owlets join those of the gourmands that are feasting while on the wing.

The earth is now green and sweet. The sugar-cane grows apace. The rice, the various millets and the other autumn crops are being sown. The cultivators take full advantage of every break in the rains to conduct agricultural operations.

As we have seen, the nocturnal chorus of the birds is now replaced by the croaking of frogs and the stridulation of crickets. In the daytime the birds still have plenty to say for themselves. The brain-fever birds scream as lustily as they did in May and June. The koel is, if possible, more vociferous than ever, especially at the beginning of the month. The Indian cuckoo does not call so frequently as formerly, but, by way of compensation, the pied crested cuckoo uplifts his voice at short intervals.

The *whoot, whoot, whoot* of the crow-pheasant booms from almost every thicket. The iora, the coppersmith, the barbet, the golden-backed woodpecker, and the white-breasted kingfisher continue to call merrily. The pied starlings are in full voice; their notes form a very pleasing addition to the avian chorus. Those magpie-robins that have not brought nesting operations to a close are singing vigorously. The king-crows are feeding their young ones in the greenwood tree, and crooning softly to them *pitchu-wee*. At the *jhils* the various waterfowl are nesting and each one proclaims the fact by its allotted call. Much strange music emanates from the well-filled tank; the indescribable cries of the purple coots, the curious "fixed bayonets" of the cotton teal and the weird cat-like mews of the jacanas form the dominant notes of the aquatic symphony.

In July the black-breasted or rain-quail (*Coturnix coromandelica*) is plentiful in India. Much remains to be discovered regarding the movements of this species. It appears to migrate to Bengal, the United Provinces, the Punjab and Sind shortly before the monsoon bursts, but it is said to arrive in Nepal as early as April. It would seem to winter in South India. It is a smaller bird than the ordinary grey quail and has no pale cross-bars on the primary wing feathers. The males of this species are held in high esteem by Indians as fighting birds. Large numbers of them are netted in the same way as the grey quail. Some captive birds are set down in a covered cage by a sugar-cane field in the evening. Their calls attract a number of wild birds, which settle down in the sugar-cane in order to spend the day there. At dawn a net is quietly stretched across one end of the field. A rope is then slowly dragged along over the growing crop in the direction of the net. This sends all the quail into the net.

Very fair sport may be obtained in July by shooting rain-quail that have been attracted by call birds.

July marks the end of one breeding season and the beginning of another. As regards the nesting season, birds fall into four classes. There is the very large class that nests in spring and summer. Next in importance is the not inconsiderable body that rears up its broods in the rains when the food supply is most abundant. Then comes the small company that builds nests in the pleasant winter time. Lastly there are the perennials—such birds as the sparrow and the dove, which nest at all seasons. In the present month the last of the summer nesting birds close operations for the year, and the monsoon birds begin to lay their eggs. July is therefore a favourable month for bird-nesting. Moreover, the sun is sometimes obscured by cloud and, under such conditions, a human being is able to remain out of doors throughout the day without suffering much physical discomfort.

With July ends the normal breeding season of the tree-pies, white-eyes, ioras; king-crows, bank-mynas, paradise flycatchers, brown rock-chats, Indian robins, dhayals, red-winged bush-larks, sunbirds, rollers, swifts, green pigeons, lapwings and butcher-birds.

The paradise flycatchers leave Northern India and migrate southwards a few weeks after the young birds have left the nest.

Numbers of bulbuls' nests are likely to be found in July, but the breeding time of these birds is rapidly drawing to its close. Sparrows and doves are of course engaged in parental duties; their eggs have been taken in every month of the year.

The nesting season is now at its height for the white-necked storks, the koels and their dupes—the house-crows, also for the various babblers and their deceivers—the brain-fever birds and the pied crested cuckoos. The tailor-birds, the ashy and the Indian wren-warblers, the brahminy mynas, the wire-tailed swallows, the amadavats, the sirkeer cuckoos, the pea-fowl, the water-hens, the common and the pied mynas, the cuckoo-shrikes and the orioles are all fully occupied with nursery duties. The earliest of the brain-fever birds to be hatched have left the nest. Like all its family the young hawk-cuckoo has a healthy appetite. In order to satisfy it the unfortunate foster-parents have to work like slaves, and often must they wonder why nature has given them so voracious a child. When it sees a babbler approaching with food, the cuckoo cries out and flaps its wings vigorously. Sometimes these completely envelop the parent bird while it is thrusting food into the yellow mouth of the cuckoo. The breast of the newly-fledged brain-fever bird is covered with dark brown drops, so that, when seen from below, it looks like a thrush with yellow legs. Its cries, however, are not at all thrushlike.

Many of the wire-tailed swallows, minivets and white-browed fantail flycatchers bring up a second brood during the rains. The loud cheerful call of the last is heard very frequently in July.

Numbers of young bee-eaters are to be seen hawking at insects; they are distinguishable from adults by the dullness of the plumage and the fact that the median tail feathers are not prolonged as bristles.

Very few crows emerge from the egg before the 1st of July, but, during the last week in June, numbers of baby koels are hatched out. The period of incubation for the koel's egg is shorter than that of the crow, hence at the outset the baby koel steals a march on his foster-brothers. Koel nestlings, when they first emerge from the egg, differ greatly in appearance from baby crows. The skin of the koel is black, that of crow is pink for the first two days of its existence, but it grows darker rapidly. The baby crow is the bigger bird and has a larger mouth with fleshy sides. The sides of the mouth of the young koel are not fleshy. The neck of the crow nestling is long and the head hangs down, whereas the koel's neck is short and the bird carries its head huddled in its shoulders. Crows nest high up in trees, these facts are therefore best observed by sending up an expert climber with a tin half-full of sawdust to which a long string is attached. The climber lets down the eggs or nestlings in the tin and the observer can examine them in comfort on *terra firma*. The parent crows do not appear to notice how unlike the young koels are to their own nestlings, for they feed them most assiduously and make a great uproar when the koels are taken from the nest. Baby crows are noisy creatures; koels are quiet and timid at first, but become noisier as they grow older.

The feathers of crow nestlings are black in each sex. Young koels fall into three classes: those of which the feathers are all black, those of which a few feathers have white or reddish tips, those which are speckled black and white all over because each feather has a white tip. The two former appear to be young cocks and the last to be hens. Baby koels, in addition to hatching out before their foster-brethren, develop more quickly, so that they leave the nest fully a week in advance of the young corvi. After vacating the nest they squat for some days on a branch close by; numbers of them are to be seen thus in suitable localities towards the end of July. At first the call of the koel is a squeak, but later it takes the form of a creditable, if ludicrous, attempt at a caw. The young cuckoo does not seem to be able to distinguish its foster-parents from other crows; it clamours for food whenever any crow comes near it.

Of the scenes characteristic of the rains in India none is more pleasing than that presented by a colony of nest-building bayas or weaver-birds (*Ploceus baya*). These birds build in company. Sometimes more than twenty of their wonderful retort-like nests are to be seen in one tree. This means that more than forty birds are at work, and, as each of these indulges in much cheerful twittering, the tree in question presents an animated scene. Both sexes take part in nest-construction.

Having selected the branch of a tree from which the nest will hang, the birds proceed to collect material. Each completed nest contains many yards of fibre not much thicker than stout thread. Such material is not found in quantity in nature. The bayas have, therefore, to manufacture it. This is easily done. The building weaver-bird betakes itself to a clump of elephant-grass, and, perching on one of the blades, makes a notch in another near the base. Then, grasping with its beak the edge of this blade above the notch, the baya flies away and thus strips off a narrow strand. Sometimes the strand adheres to the main part of the blade at the tip so firmly that the force of the flying baya is not sufficient to sever it. The bird then swings for a few seconds in mid-air, suspended by the strip of leaf. Not in the least daunted the baya makes a fresh effort and flies off, still gripping the strand firmly. At the third, if not at the second attempt, the thin strip is completely severed. Having secured its prize the weaver-bird proceeds to tear off one or two more strands and then flies with these in its bill to the nesting site, uttering cries of delight. The fibres obtained in this manner are bound round the branch from which the nest will hang. More strands are added to form a stalk; when this has attained a length of several inches it is gradually expanded in the form of an umbrella or bell. The next step is to weave a band of grass across the mouth of the bell. In this condition the nest is often left unfinished. Indians call such incomplete nests *jhulas* or swings; they assert that these are made in order that the cocks may sit in them and sing to their mates while these are incubating the eggs. It may be, as "Eha" suggests, that at this stage the birds are dissatisfied with the balance of the nest and for this reason leave it. If the nest, at this point of its construction, please the weaver-birds they proceed to finish it by closing up the bell at one side of the cross-band to form a receptacle for the eggs, and prolonging the other half of the bell into a long tunnel or neck. This neck forms the entrance to the nest; towards its extremity it becomes very flimsy so that it affords no foothold to an enemy. Nearly every baya's nest contains some lumps of clay attached to it. Jerdon was of opinion that the function of these is to balance the nest properly. Indians state that the bird sticks fireflies into the lumps of clay to light up the nest at night. This story has found its way into some ornithological text-books. There is no truth in it. The present writer is inclined to think that the object of these lumps of clay is to prevent the light loofah-like nest swinging too violently in a gale of wind.

Both sexes take part in nest-construction. After the formation of the cross-bar at the mouth of the bell one of the birds sits inside and the other outside, and they pass the strands to each other and thus the weaving proceeds rapidly. While working at the nest the bayas, more especially the cocks, are in a most excited state. They sing, scream, flap their wings and snap the bill. Sometimes one cock in his excitement attacks a neighbour by jumping on his back! This results in a fight in which the birds flutter in the air, pecking at one another. Often the combatants "close" for a few seconds, but neither bird seems to get hurt in these little contests.

Every bird-lover should make a point of watching a company of weaver-birds while these are constructing their nests. The tree or trees in which they build can easily be located by sending a servant in July to search for them. The favourite sites for nests in the United Provinces seem to be babul trees that grow near borrow pits alongside the railroad.

In the rainy season two other birds weave nests, which are nearly as elegant as those woven by the baya. These birds, however, do not nest in company. They usually build inside bushes, or in long grass.

For this reason they do not lend themselves to observation while at work so readily as bayas do. The birds in question are the Indian and the ashy wren-warbler.

The former species brings up two broods in the year. One, as has been mentioned, in March and the other in the "rains."

The nest of the Indian wren-warbler (*Prinia inornata*) is, except for its shape and its smaller size, very like that of a weaver-bird. It is an elongated purse or pocket, closely and compactly woven with fine strips of grass from 1/40 to 1/20 inch in breadth. The nest is entered by a hole near the top. Both birds work at the nest, clinging first to the neighbouring stems of grass or twigs, and later to the nest itself when this has attained sufficient dimensions to afford them foothold. They push the ends of the grass in and out just as weaver-birds do. Like the baya, the Indian wren-warbler does not line its nest. The eggs are pale greenish-blue, richly marked by various shades of deep chocolate and reddish-brown. As Hume remarks: "nothing can exceed the beauty or variety of markings, which are a combination of bold blotches, clouds and spots, with delicate, intricately woven lines, recalling somewhat ... those of our early favourite—the yellow-hammer."

The ashy wren-warbler (*Prinia socialis*) builds two distinct kinds of nest. One is just like that of the tailor-bird, being formed by sewing or cobbling together two, three, four or five leaves, and lining the cup thus formed with down, wool, cotton or other soft material. The second kind of nest is a woven one. This is a hollow ball with a hole in the side. The weaving is not so neat as that of the baya and the Indian wren-warbler. Moreover, several kinds of material are usually worked into the nest, which is invariably lined.

The building of two totally different types of nest is an interesting phenomenon, and seems to indicate that under the name *Prinia socialis* are classed two different species, which anatomically are so like one another that systematists are unable to separate them. Both kinds of nests are found in the same locality and at the same time of the year. Against the theory that there are two species of ashy wren-warbler is the fact that there is no difference in appearance between the eggs found in the two kinds of nest. All eggs are brick-red or mahogany colour, without any spots or markings.

Many of the Indian cliff-swallows, of which the nests are described in the calendar for March, bring up a second brood in the "rains."

Needless to state that in the monsoon the tank and the *jhil* are the happy hunting grounds of the ornithologist.

In July and August not less than thirty species of waterfowl nidificate. Floating nests are constructed by sarus cranes, purple coots and the jacanas. The various species of egrets breed in colonies in trees in some village not far from a tank; in company with them spoonbills, cormorants, snake-birds, night-herons and other birds often nest. The white-breasted waterhen constructs its nursery in a thicket at the margin of some village pond. The resident ducks are also busy with their nests. These are in branches of trees, in holes in trees or old buildings, or on the ground.

When describing the nesting operations of waterfowl in Northern India it is difficult to apportion these between July and August, for the eggs of almost all such species are as likely to be found in the one month as in the other. A few individuals begin to lay in June, the majority commence in July, but a great many defer operations until August. There is scarcely an aquatic species of which it can be said: "It never lays before August." Nor are there many of which it can be asserted: "Their eggs are never found after July."

Individuals differ in their habit. A retarded monsoon means that the water-birds begin to nest later than usual. The first fall of the monsoon rain seems to be the signal for the commencement of nesting operations, but by no means every pair of birds obeys the signal immediately.

The nearest approach to a generalisation which it is possible to make is that the egrets and paddy-birds are usually the first of the monsoon breeders to begin nest-building, while the spot-billed duck, the whistling teal and the bronze-winged jacana are the last. In other words, the eggs of the former are most likely to be found in July and those of the latter in August.

As the calendar for this month has already attained considerable dimensions, a description of the nests of all these water-birds is given in the August calendar. It is, however, necessary to state that the eggs of the following birds are likely to be found in July: purple coot, common coot, bronze-winged and pheasant-tailed jacana, black ibis, white-necked stork, cormorant, snake-bird, cotton teal, comb duck, spot-billed duck, spoonbill, and the various herons and egrets.

AUGUST

> See! the flushed horizon flames intense
> With vivid red, in rich profusion streamed
> O'er heaven's pure arch. At once the clouds assume
> Their gayest liveries; these with silvery beams
> Fringed lovely; splendid those in liquid gold,
> And speak their sovereign's state. He comes, behold!
>
> MALLET.

The transformation scene described in July continues throughout August. Torrential rain alternates with fierce sunshine. The earth is verdant with all shades of green. Most conspicuous of these are the yellowish verdure of the newly-transplanted rice, the vivid emerald of the young plants that have taken root, the deeper hue of the growing sugar-cane, and the dark green of the mango topes.

Unless the monsoon has been unusually late in reaching Northern India the autumn crops are all sown before the first week in August. The sugar-cane is now over five feet in height. The cultivators are busily transplanting the better kinds of rice, or running the plough through fields in which the coarser varieties are growing.

The aloes are in flower. Their white spikes of drooping tulip-like flowers are almost the only inflorescences to be seen outside gardens at this season of the year. The mango crop is over, but that of the pineapples takes its place.

At night-time many of the trees are illumined by hundreds of fireflies. These do not burn their lamps continuously. Each insect lets its light shine for a few seconds and then suddenly puts it out. It sometimes happens that all the fireflies in a tree show their lights and extinguish them simultaneously and thereby produce a luminous display which is strikingly beautiful. Fireflies are to be seen during the greater part of the year, but they are far more abundant in the "rains" than at any other season.

As in July so in August the voices of the birds are rarely heard after dark. The nocturnal music is now the product of the batrachian band, ably seconded by the crickets.

During a prolonged break in the rains the frogs and toads are hushed, except in *jhils* and low-lying paddy fields. Cessation of the rain, however, does not silence the crickets.

The first streak of dawn is the signal for the striking up of the jungle and the spotted owlets. Hard upon them follow the koels and the brain-fever birds. These call only for a short time, remaining silent during the greater part of the day. Other birds that lift up their voices at early dawn are the crow-pheasant, the black partridge and the peacock. These also call towards dusk. As soon as the sun has risen the green barbets, coppersmiths, white-breasted kingfishers and king-crows utter their familiar notes; even these birds are heard but rarely in the middle of the day, nor have their voices the vigour that characterised them in the hot weather. Occasionally the brown rock-chat emits a few notes, but he does so in a half-hearted manner. In the early days of August the magpie-robins sing at times; their song, however, is no longer the brilliant performance it was. By the end of the month it has completely died away.

The Indian cuckoo no more raises its voice in the plains, but the pied crested-cuckoo continues to call lustily and the pied starlings make a joyful noise. The oriole's liquid *pee-ho* is gradually replaced by the loud *tew*, which is its usual cry at times when it is not nesting.

The water-birds, being busy at their nests, are of course noisy, but, with the exception of the loud trumpeting of the sarus cranes, their vocal efforts are heard only at the *jhil*.

The did-he-do-its, the rollers, the bee-eaters, two or three species of warblers and the perennial singers complete the avian chorus.

Numbers of rosy starlings are returning from Asia Minor, where they have reared up their broods. The inrush of these birds begins in July and continues till October. They are the forerunners of the autumn immigrants. Towards the end of the month the garganey or blue-winged teal (*Querquedula circia*), which are the earliest of the migratory ducks to visit India, appear on the tanks. Along with them comes the advance-guard of the snipe. The pintail snipe (*Gallinago stenura*) are invariably the first to appear, but they visit only the eastern parts of Northern India. Large numbers of them sojourn in Bengal and Assam. Stragglers appear in the eastern portion of the United Provinces; in the western districts and in the Punjab this snipe is a *rara avis*. By the third week in August good bags of pintail snipe are sometimes obtained in Bengal. The fantail or full-snipe (*G. coelestis*) is at least one week later in arriving. This species has been shot as early as the 24th August, but there is no general immigration of even the advance-guard until quite the end of the month.

The jack-snipe (*G. gallinula*) seems never to appear before September.

Most of the monsoon broods of the Indian cliff-swallow emerge from the eggs in August. The "rains" breeding season of the amadavats or red munias is now over, and the bird-catcher issues forth to snare them.

His stock-in-trade consists of some seed and two or three amadavats in one of the pyramid-shaped wicker cages that can be purchased for a few annas in any bazaar. To the base of one of the sides of the cage a flap is attached by a hinge. The flap, which is of the same shape and size as the side of the cage, is composed of a frame over which a small-meshed string net is stretched. A long string is fastened to the apex of the flap and passed through a loop at the top of the cage. Selecting an open space near some tall grass in which amadavats are feeding, the bird-catcher sets down the cage and loosens the string so that the flap rests on the earth. Some seed is sprinkled on the flap. Then the trapper squats behind a bush, holding the end of the string in his hand. The cheerful little *lals* inside the cage soon begin to twitter and sing, and their calls attract the wild amadavats in the vicinity. These come to the cage, alight on the flap, and begin to eat the seed. The bird-catcher gives the string a sharp pull and thus traps his victims between the flap and the side of the cage. He then disentangles them, places them in the cage, and again sets the trap.

Almost all the birds that rear up their young in the spring have finished nesting duties for the year by August. Here and there a pair of belated rollers may be seen feeding their young. Before the beginning of the month nearly all the young crows and koels have emerged from the egg, and the great majority of them have left the nest. Young house-crows are distinguished from adults by the indistinctness of the grey on the neck. They continually open their great red mouths to clamour for food.

The wire-tailed swallows, swifts, pied crested-cuckoos, crow-pheasants, butcher-birds, cuckoo-shrikes, fantail flycatchers, babblers, white-necked storks, wren-warblers, weaver-birds, common and pied mynas, peafowl, and almost all the resident water-birds, waders and swimmers, except the terns and the plovers, are likely to have eggs or young. The nesting season of the swifts and butcher-birds is nearly over. In the case of the others it is at its height. The wire-tailed swallows and minivets are busy with their second broods. The nests of most of these birds have already been described.

The Indian peafowl (*Pavo cristatus*) usually lay their large white eggs on the ground in long grass or thick undergrowth. Sometimes they nestle on the grass-grown roofs of deserted buildings or in other elevated situations. Egrets, night-herons, cormorants, darters, paddy-birds, openbills, and spoonbills build stick nests in trees. These birds often breed in large colonies. In most cases the site chosen is a clump of trees in a village which is situated on the border of a tank. Sometimes all these species nest in company. Hume described a village in Mainpuri where scores of the above-mentioned birds, together with some whistling teal and comb-ducks, nested simultaneously. After a site has been selected by a colony the birds return year after year to the place for nesting purposes. The majority of the eggs are laid in July, the young appearing towards the end of that month or early in the present one.

The nest of the sarus crane (*Grus antigone*) is nearly always an islet some four feet in diameter, which either floats in shallow water or rises from the ground and projects about a foot above the level of the water. The nest is composed of dried rushes. It may be placed in a *jhil*, a paddy field, or a borrow pit by the railway line. A favourite place is the midst of paddy cultivation in some low-lying field where the water is too deep to admit of the growing of rice. Two very large white eggs, rarely three, are laid. This species makes no attempt to conceal its nest. In the course of a railway journey in

August numbers of incubating saruses may be seen by any person who takes the trouble to look for them.

"Raoul" makes the extraordinary statement that incubating sarus cranes do not sit when incubating, but hatch the eggs by standing over them, one leg on each side of the nest! Needless to say there is no truth whatever in this statement. The legs of the sitting sarus crane are folded under it, as are those of incubating flamingos and other long-legged birds.

Throughout the month of August two of the most interesting birds in India are busy with their nests. They are the pheasant-tailed and the bronze-winged jacana. These birds live, move and have their being on the surface of lotus-covered tanks. Owing to the great length of their toes jacanas are able to run about with ease over the surface of the floating leaves of water-lilies and other aquatic plants, or over tangled masses of rushes and water-weeds.

In the monsoon many tanks are so completely covered with vegetation that almost the only water visible to a person standing on the bank consists of the numerous drops that have been thrown on to the flat surfaces of the leaves, where they glisten in the sun like pearls.

Two species of jacana occur in India: the bronze-winged (*Motopus indicus*) and the pheasant-tailed jacana or the water-pheasant (*Hydrophasianus chirurgus*). They are to be found on most tanks in the well-watered parts of the United Provinces. They occur in small flocks and are often put up by sportsmen when shooting duck. They emit weird mewing cries. The bronze-winged jacana is a black bird with bronze wings. It is about the size of a pigeon, but has much longer legs. The pheasant-tailed species is a black-and-white bird. In winter the tail is short, but in May both sexes grow long pheasant-like caudal feathers which give the bird its popular name. The bronze-winged jacana does not grow these long tail feathers.

The nests of jacanas are truly wonderful structures. They are just floating pads of rushes and leaves of aquatic plants. Sometimes practically the whole of the pad is under water, so that the eggs appear to be resting on the surface of the tank. The nest of the bronze-winged species is usually larger and more massive than that of the water-pheasant. The latter's nest is sometimes so small as hardly to be able to contain the eggs—a little, shallow, circular cup of rushes and water-weeds or floating lotus leaves or tufts of water-grass. The eggs of the two species show but little similarity. Both, however, are very beautiful and remarkable. The eggs of the bronze-winged jacana have a rich brownish-bronze background, on which black lines are scribbled in inextricable confusion, so that the egg looks as though Arabic texts had been scrawled over it. This species might well be called "the Arabic writing-master." The eggs of the water-pheasant are in shape like pegtops without the peg. They are of a dark rich green-bronze colour, and devoid of any markings.

The nest of the handsome, but noisy, purple coot (*Porphyrio poliocephalus*) is a platform of rushes and reeds which is sometimes placed on the ground in a rice field, but is more often floating, and is then tethered to a tree or some other object. From six to ten eggs are laid. These are very beautiful objects. The ground colour is delicate pink. This is spotted and blotched with crimson; beneath these spots there are clouds of pale purple which have the appearance of lying beneath the surface of the shell.

The white-breasted water-hen (*Gallinula phoenicura*) is a bird that must be familiar to all. One pair, at least, is to be found in every village which boasts of a tank and a bamboo clump, no matter how small these be. The water-hen is a black bird about the size of the average bazaar fowl, with a white face, throat and breast. It carries its short tail almost erect, and under this is a patch of brick-red feathers. During most seasons of the year it is a silent bird, but from mid-May until the end of the monsoon it is exceedingly noisy, and, were it in the habit of haunting our gardens and compounds, its cries would attract as much attention as do those of the koel and the brain-fever bird. As, however, water-hens are confined to tiny hamlets situated far away from cities, many people are not acquainted with their calls, which "Eha" describes as "roars, hiccups and cackles." The nest is built in a bamboo clump or other dense thicket. The eggs are stone-coloured, with spots of brown, red and purple. The young birds, when first hatched, are covered with black down, and look like little black ducklings. They can run, swim and dive as soon as they leave the egg. Little parties of them are to be seen at the edge of most village tanks in August.

The resident ducks are all busy with their nests. The majority of them lay their eggs in July, so that in August they are occupied with their young.

The cotton-teal (*Nettopus coromandelianus*) usually lays its eggs in a hole in a mango or other tree. The hollow is sometimes lined with feathers and twigs. It is not very high up as a rule, from six to twelve feet above the ground being the usual level. The tree selected for the nesting site is not necessarily close to water. Thirteen or fourteen eggs seem to be the usual clutch, but as many as twenty-two have been taken from one nest. Young teal, when they emerge from the egg, can swim and walk, but they are unable to fly. No European seems to have actually observed the process whereby they get from the nest to the ground or the water. It is generally believed that the parent birds carry them. Mr. Stuart Baker writes that a very intelligent native once told him that, early one morning, before it was light, he was fishing in a tank, when he saw a bird flutter heavily into the water from a tree in front of him and some twenty paces distant. The bird returned to the tree, and again, with much beating of the wings, fluttered down to the surface of the tank; this performance was repeated again and again at intervals of some minutes. At first the native could only make out that the cause of the commotion was a bird of some kind, but after a few minutes, he, remaining crouched among the reeds and bushes, saw distinctly that it was a cotton-teal, and that each time it flopped into the water and rose again it left a gosling behind it. The young ones were carried somehow in the feet, but the parent bird seemed to find the carriage of its offspring no easy matter; it flew with difficulty, and fell into the water with considerable force.

August is the month in which some fortunate observer will one year be able to confirm or refute this story.

The comb-duck or *nukta* (*Sarcidiornis melanotus*), which looks more like a freak of some domesticated breed than one of nature's own creatures, makes, in July or August, a nest of grass and sticks in a hole in a tree or in the fork of a stout branch. Sometimes disused nests of other species are utilised. About a dozen eggs is the usual number of the clutch, but Anderson once found a nest containing no fewer than forty eggs.

The lesser whistling-teal (*Dendrocygna javanica*) usually builds its nest in a hollow in a tree. Sometimes it makes use of the deserted nursery of another species, and there are many cases on record of the nest being on the ground, a *bund*, or a piece of high ground in a *jhil*. Eight or ten eggs are laid.

The little grebe or dabchick (*Podiceps albipennis*) is another species that lays in July or August. This bird, which looks like a miniature greyish-brown duck without a tail, must be familiar to Anglo-Indians, since at least one pair are to be seen on almost every pond or tank in Northern India. Although permanent residents in this country, little grebes leave, in the "rains," those tanks that do not afford plenty of cover, and betake themselves to a *jhil* where vegetation is luxuriant. The nest, like that of other species that build floating cradles, is a tangle of weeds and rushes. When the incubating bird leaves the nest she invariably covers the white eggs with wet weeds, and, as Hume remarks, it is almost impossible to catch the old bird on the nest or to take her so much by surprise as not to allow her time to cover up the eggs. As a matter of fact, these birds spend very little time upon the nest in the day-time. The sun's rays are powerful enough not only to supply the heat necessary for incubation but to bake the eggs. This *contretemps*, however, is avoided by placing wet weeds on the eggs and by the general moisture of the nest. No better idea of the heat of India during the monsoon can be furnished than that afforded by the case of some cattle-egrets' eggs taken by a friend of the writer's in August, 1913. He found a clutch of four eggs; not having leisure at the time to blow them, he placed them in a bowl on the drawing-room mantelshelf. On the evening of the following day he heard some squeaks, but, thinking that these sounds emanated from a musk-rat or one of the other numerous rent-free tenants of every Indian bungalow, paid little heed to them. When, however, the same sounds were heard some hours later and appeared to emanate from the mantelpiece, he went to the bowl, and, lo and behold, two young egrets had emerged! These were at once fed. They lived for three days and appeared to be in good health, when they suddenly gave up the ghost.

SEPTEMBER

And sweet it is by lonely meres
To sit, with heart and soul awake,
Where water-lilies lie afloat,
Each anchored like a fairy boat
Amid some fabled elfin lake:
To see the birds flit to and fro
Along the dark-green reedy edge.

MARY HOWITT.

September is a much-abused month. Many people assert that it is the most unpleasant and unhealthy season of the year.

Malarial and muggy though it is, September scarcely merits all the evil epithets that are applied to it. The truth is that, after the torrid days of the hot weather and the humid heat of the rainy season, the European is thoroughly weary of his tropical surroundings, his vitality is at a low ebb, he is languid and irritable, thus he complains bitterly of the climate of September, notwithstanding the fact that it is a distinct improvement on that of the two preceding months.

In the early part of the month the weather differs little from that of July and August. The days are somewhat shorter and the sun's rays somewhat less powerful, in consequence the average temperature is slightly lower. Normally the rains cease in the second half of the month. Then the sky resumes the fleckless blueness which characterises it during the greater part of the year. The blue of the sky is more pure and more intense in September than at other times, except during breaks in the monsoon, because the rain has washed from the atmosphere the myriads of specks of dust that are usually suspended in it.

The cessation of the rains is followed by a period of steamy heat. As the moisture of the air gradually diminishes the temperature rises. But each September day is shorter than the one before it, and, hour by hour, the rays of the sun part with some of their power. Towards the end of the month the nights are cooler than they have been for some time. At sunset the village smoke begins to hang low in a diaphanous cloud—a sure sign of the approaching cold weather. The night dews are heavy. In the morning the blades of grass and the webs of the spiders are bespangled with pearly dewdrops. Cool zephyrs greet the rising sun. At dawn there is, in the last days of the month, a touch of cold in the air.

The Indian countryside displays a greenness which is almost spring-like; not quite spring-like, because the fierce greens induced by the monsoon rains are not of the same hues as those of the young leaves of spring. The foliage is almost entirely free from dust. This fact adds to the vernal appearance of the landscape. The _jhils_ and tanks are filled with water, and, being overgrown with luxuriant vegetation, enhance the beauty of the scene. But, almost immediately after the cessation of the rains, the country begins to assume its usual look. Day by day the grass loses a little of its greenness. The earth dries up gradually, and its surface once more becomes dusty. The dust is carried to the foliage, on which it settles, subduing the natural greenery of the leaves. No sooner do the rains cease than the rivers begin to fall. By November most of them will be sandy wastes in which the insignificant stream is almost lost to view.

The mimosas flower in September. Their yellow spherical blossoms are rendered pale by contrast with the deep gold hue of the blooms of the _san_ (hemp) which now form a conspicuous feature of the landscape in many districts. The cork trees (_Millingtonia hortensis_) become bespangled with hanging clusters of white, long-tubed, star-like flowers that give out fragrant perfume at night.

The first-fruits of the autumn harvest are being gathered in. Acre upon acre of the early-sown rice falls before the sickle. The threshing-floors once again become the scene of animation. The fallow fields are being prepared for the spring crops and the sowing of the grain is beginning.

Throughout the month insect life is as rich and varied as it was in July and August.

The brain-fever bird and the koel call so seldom in September that their cries, when heard, cause surprise. The voice of the pied crested-cuckoo no longer falls upon the ear, nor does the song of the magpie-robin. The green barbets lift up their voices fairly frequently, but it is only on rare occasions that their cousins—the coppersmiths—hammer on their anvils. The pied mynas are far less vociferous than they were in July and August.

By the end of September the bird chorus has assumed its winter form, except that the grey-headed flycatchers have not joined it in numbers.

Apart from the sharp notes of the warblers, the cooing of the doves, the hooting of the crow-pheasants, the wailing of the kites, the cawing of the crows, the screaming of the green parrots, the chattering of the mynas and the seven sisters, the trumpeting of the sarus cranes and the clamouring of the lapwings, almost the only bird voices commonly heard are those of the fantail flycatcher, the amadavat, the wagtail, the oriole, the roller and the sunbird.

The cock sunbirds are singing brilliantly although they are still wearing their workaday garments, which are quaker brown save for one purple streak along the median line of the breast and abdomen.

Many birds are beginning to moult. They are casting off worn feathers and assuming the new ones that will keep them warm during the cool winter months. With most birds the new feathers grow as fast as the old ones fall out. In a few, however, the process of renewal does not keep pace with that of shedding; the result is that the moulting bird presents a mangy appearance. The mynas afford conspicuous examples of this; when moulting their necks often become almost nude, so that the birds bear some resemblance to miniature vultures.

Great changes in the avifauna take place in September.

The yellow-throated sparrows, the koels, the sunbirds, the bee-eaters, the red turtle-doves and the majority of the king-crows leave the Punjab. From the United Provinces there is a large exodus of brain-fever birds, koels, pied crested-cuckoos, paradise flycatchers and Indian orioles. These last are replaced by black-headed orioles in the United Provinces, but not in the Punjab.

On the other hand, the great autumnal immigration takes place throughout the month. Before September is half over the migratory wagtails begin to appear. Like most birds they travel by night when migrating. They arrive in silence, but on the morning of their coming the observer cannot fail to notice their cheerful little notes, which, like the hanging of the village smoke, are to be numbered among the signs of the approach of winter. The three species that visit India in the largest numbers are the white (*Motacilla alba*), the masked (*M. personata*) and the grey wagtail (*M. melanope*). In Bengal the first two are largely replaced by the white-faced wagtail (*M. leucopsis*). The names "white" and "grey" are not very happy ones. The white species is a grey bird with a white face and some black on the head and breast; the masked wagtail is very difficult to distinguish from the white species, differing in having less white and more black on the head and face, the white constituting the "mask"; the grey wagtail has the upper plumage greenish-grey and the lower parts sulphur-yellow. The three species arrive almost simultaneously, but the experience of the writer is that the grey bird usually comes a day or two before his cousins.

On one of the last ten days of September the first batch of Indian redstarts (*Ruticilla frontalis*) reaches India. Within twenty days of the coming of these welcome little birds it is possible to dispense with punkas.

Like the redstarts the rose-finches and minivets begin to pour into India towards the end of September. The snipe arrive daily throughout the month.

With the first full moon of September come the grey quail (*Coturnix communis*). These, like the rain-quail, afford good sport with the gun if attracted by call birds set down overnight. When the stream of immigrating quail has ceased to flow, these birds spread themselves over the well-cropped country. It then becomes difficult to obtain a good bag of quail until the time of the spring harvest, when they collect in the crops that are still standing.

Thousands of blue-winged teal invade India in September, but most of the other species of non-resident duck do not arrive until October or even November.

Not the least important of the September arrivals are the migratory birds of prey. None of the owls seem to migrate. Nor do the vultures, but a large proportion of the diurnal raptores leaves the plains of India in the spring.

To every migratory species of raptorial bird, that captures living quarry, there is a non-migratory counterpart or near relative. It would almost seem as if each species were broken up into two clans—a migratory and a stationary one. Thus, of each of the following pairs of birds the first-named is migratory and the other non-migratory: the steppe-eagle and the tawny eagle, the large Indian and the common kite, the long-legged and the white-eyed buzzard, the sparrow-hawk and the shikra, the peregrine and the lugger falcon, the common and the red-headed merlin, the kestrel and the black-winged kite.

It is tempting to formulate the theory that the raptores are migratory or the reverse according or not as they prey on birds of passage, and that the former migrate merely in order to follow their quarry. Certain facts seem to bear out this theory. The peregrine falcon, which feeds largely on ducks, is migratory, while the lugger falcon—a bird not particularly addicted to waterfowl—remains in India throughout the year.

The necessity of following their favourite quarry may account for the migratory habits of some birds of prey, but it does not apply to all. Thus, the osprey, which feeds almost exclusively on fish, is merely a winter visitor to India. Again, there is the kestrel. This preys on non-migratory rats and mice, nevertheless it leaves the plains in the hot weather and goes to the Himalayas to breed. All the species of birds of prey cited above as migratory begin to arrive in the plains of India in September. The merlins come only into the Punjab, but most of the other raptores spread over the whole of India.

The various species of harrier make their appearance in September. These are birds that cannot fail to attract attention. They usually fly slowly a few feet above the surface of the earth so that they can drop suddenly on their quarry. They squat on the ground when resting, but their wings are long and their bodies light, so that they do not need much rest. Those who shoot duck have occasion often to say hard things of the marsh-harrier and the peregrine falcon, because these birds are apt to come as unbidden guests to the shoot and carry off wounded duck and teal before the *shikari* has time to retrieve them.

Of the migratory birds of prey the kestrel is perhaps the first to arrive; the osprey and the peregrine falcon are among the last.

Very few observations of the comings and the goings of the various raptorial birds have been recorded; in the present state of our knowledge it is not possible to compile an accurate table showing the usual order in which the various species appear. This is a subject to which those persons who dwell permanently in one place might with advantage direct their attention.

As regards nesting operations September is not a month of activity.

On the 15th the close season for game birds ends in the Government forests; and by that date the great majority of them have reared up their broods. Grey partridge's eggs, it is true, have been taken in September; but as we have seen, grey partridges, like doves and kites, can scarcely be said to have a breeding season; they lay eggs whenever it seemeth good to them.

A few belated peafowl may still be found with eggs, but these are exceptions. Most of the hens are strutting about proudly, accompanied by their chicks, while the cocks are shedding their trains. Other species of which the eggs may be found in the present month are the white-throated munia, the common and the large grey babblers, and, of course, the various species of dove.

Before the last day of August all the young mynas have emerged from the egg, and throughout the first half of September numbers of them are to be seen following their parents and clamouring for food. Most of the koels have departed, but some individuals belonging to the rising generation remind us that they are still with us by emitting sounds which are very fair imitations of the "sqwaking" of young crows.

Baby koels are as importunate as professional beggars and solicit food of every crow that passes by, to the great disgust of all but their foster-parents.

The majority of the seven sisters have done with nursery duties for a season. Some flocks, however, are still accompanied by impedimenta in the shape of young babblers or pied crested-cuckoos. The impedimenta make far more noise than the adult birds. They are always hungry, or at any rate always demanding food in squeaky tones. With each squeak the wings are flapped violently, as if to emphasise the demand. Every member of a flock appears to help to feed the young birds irrespective of whose nests these have been reared in.

Throughout September bayas are to be seen at their nests, but, before the month draws to its close, nearly all the broods have come out into the great world. The nests will remain until next monsoon, or even longer, as monuments of sound workmanship.

In September numbers of curious brown birds, heavily barred with black, make their appearance. These are crow-pheasants that have emerged from nests hidden away in dense thickets. In a few weeks these birds will lose their barred feathers and assume the black plumage and red wings of the adult. By the end of August most of the night-herons and those of the various species of egrets that have not been killed by the plume-hunters are able to congratulate themselves on having successfully reared up their broods. In September they lose their nuptial plumes.

OCTOBER

> Ye strangers, banished from your native glades,
> Where tyrant frost with famine leag'd proclaims
> "Who lingers dies"; with many a risk ye win
> The privilege to breathe our softer air
> And glean our sylvan berries.
>
> GISBORNE'S *Walks in a Forest.*

October in India differs from the English month in almost every respect. The one point of resemblance is that both are periods of falling temperature.

In England autumn is the season for the departure of the migratory birds; in India it is the time of their arrival.

The chief feature of the English October—the falling of the leaves—is altogether wanting in the Indian autumn.

Spring is the season in which the pulse of life beats most vigorously both in Europe and in Asia; it is therefore at that time of year that the trees renew their garments.

In England leaves are short-lived. After an existence of about six months they "curl up, become brown, and flutter from their sprays." In India they enjoy longer lives, and retain their greenness for the greater part of a year. A few Indian trees, as, for example, the shesham, lose their foliage in autumn; the silk-cotton and the coral trees part with their leaves gradually during the early months of the winter, but these are the exceptions; nearly all the trees retain their old leaves until the new ones appear in spring, so that, in this country, March, April and May are the months in which the dead leaves lie thick upon the ground.

In many ways the autumn season in Northern India resembles the English spring. The Indian October may be likened to April in England. Both are months of hope, heralds of the most pleasant period of the year. In both the countryside is fresh and green. In both millions of avian visitors arrive.

Like the English April, October in Northern India is welcome chiefly for that to which it leads. But it has merits of its own. Is not each of its days cooler than the preceding one? Does it not produce the joyous morn on which human beings awake to find that the hot weather is a thing of the past?

Throughout October the sun's rays are hot, but, for an hour or two after dawn, especially in the latter half of the month, the climate leaves little to be desired. An outing in the early morning is a thing of joy, if it be taken while yet the air retains the freshness imparted to it by the night, and before the grass has yielded up the sparkling jewels acquired during the hours of darkness. It is good to ride forth on an October morn with the object of renewing acquaintance with nimble wagtails, sprightly redstarts, stately demoiselle cranes and other newly-returned migrants. In addition to meeting many winter visitors, the rider may, if he be fortunate, come upon a colony of sand-martins that has begun nesting operations.

The husbandman enjoys very little leisure at this season of the year. From dawn till sunset he ploughs, or sows, or reaps, or threshes, or winnows.

The early-sown rice yields the first-fruits of the _kharif_ harvest. By the end of the month it has disappeared before the sickle and many of the fields occupied by it have been sown with gram. The hemp (*san*) is the next crop to mature. In some parts of Northern India its vivid yellow flowers are the most conspicuous feature of the autumn landscape. They are as brilliantly coloured as broom. The *san* plant is not allowed to display its gilded blooms for long, it is cut down in the prime of life and cast into a village pond, there to soak. The harvesting of the various millets, the picking of the cotton, and the sowing of the wheat, barley, gram and poppy begin before the close of the month. The sugar-cane, the _arhar_ and the late-sown rice are not yet ready for the sickle. Those crops will be cut in November and December.

As in September so in October the birds are less vociferous than they were in the spring and the hot weather. During the earlier part of the month the notes of the koel and the brain-fever bird are heard on rare occasions; before October has given place to November, these noisy birds cease to trouble. The pied starlings have become comparatively subdued, their joyful melody is no longer a notable feature of the avian chorus. In the first half of the month the green barbets utter their familiar cries at frequent intervals; as the weather grows colder they call less often, but at no season of the year do they cease altogether to raise their voices. The *tonk, tonk, tonk* of the coppersmith is rarely heard in October; during the greater part of the cold weather this barbet is a silent creature, reminding us of its presence now and then by calling out *wow* softly, as if half ashamed at the sound of its voice. The oriole now utters its winter note *tew*, and that sound is heard only occasionally.

It is unnecessary to state that the perennials—the crows, kites, doves, bee-eaters, tree-pies, tailor-birds, cuckoo-shrikes, green parrots, jungle and spotted owlets—are noisy throughout the month.

The king-crows no longer utter the soft notes which they seem to keep for the rainy season; but, before settling down to the sober delights of the winter, some individuals become almost as lively and vociferous as they were in the nesting season. Likewise some pairs of "blue jays" behave, in September and October, as though they were about to recommence courtship; they perform strange evolutions in the air and emit harsh cries, but these lead to nothing; after a few days of noisy behaviour the birds resume their more normal habits.

The hoopoes have been silent for some time, but in October a few of them take up their refrain—*uk-uk-uk-uk*, and utter it with almost as much vigour as they did in March.

It would thus seem that the change of season, the approach of winter, has a stimulating influence on king-crows, rollers and hoopoes, causing the energy latent within them suddenly to become active and to manifest itself in the form of song or dance.

In October the pied chat and the wood-shrike frequently make sweet melody. Throughout the month the cock sunbirds sing as lustily and almost as brilliantly as canaries; many of them are beginning to reassume the iridescent purple plumage which they doffed some time ago. From every mango tope emanates the cheerful lay of the fantail flycatcher and the lively "Think of me ...

Never to be" of the grey-headed flycatcher. Amadavats sing sweet little songs without words as they flit about among the tall grasses.

In the early morning and at eventide, the crow-pheasants give vent to their owl-like hoot, preceded by a curious guttural *kok-kok-kok*. The young ones, that left the nest some weeks ago, are rapidly losing their barred plumage and are assuming the appearance of the adult. By the middle of November very few immature crow-pheasants are seen.

Migration and moulting are the chief events in the feathered world at the present season. The flood of autumn immigration, which arose as a tiny stream in August, and increased in volume nightly throughout September, becomes, in October, a mighty river on the bosom of which millions of birds are borne.

Day by day the avian population of the *jhils* increases. At the beginning of the month the garganey teal are almost the only migratory ducks to be seen on them. By the first of November brahminy duck, gadwall, common teal, widgeon, shovellers and the various species of pochard abound. With the duck come demoiselle cranes, curlews, storks, and sandpipers of various species. The geese and the pintail ducks, however, do not return to India until November. These are the last of the regular winter visitors to come and the first to go.

The various kinds of birds of prey which began to appear in September continue to arrive throughout the present month.

Grey-headed and red-breasted flycatchers, minivets, bush-chats, rose-finches and swallows pour into the plains from the Himalayas, while from beyond those mountains come redstarts, wagtails, starlings, buntings, blue-throats, quail and snipe. Along with the other migrants come numbers of rooks and jackdaws. These do not venture far into India; they confine themselves to the North-West Frontier Province and the Punjab, where they remain during the greater part of the winter. The exodus, from the above-mentioned Provinces, of the bee-eaters, sunbirds, yellow-throated sparrows, orioles, red turtle-doves and paradise flycatchers is complete by the end of October. The above are by no means the only birds that undergo local migration. The great majority of species probably move about in a methodical manner in the course of the year; a great deal of local migration is overlooked, because the birds that move away from a locality are replaced by others of their kind that come from other places.

During a spell of exceptionally cold weather a great many Himalayan birds are driven by the snow into the plains of India, where they remain for a few days or weeks. Some of these migrants are noticed in the calendar for December.

In October the annual moult of the birds is completed, so that, clothed in their warm new feathers, they are ready for winter some time before it comes. In the case of the redstart, the bush-chat, most of the wagtails, and some other species, the moult completely changes the colouring of the bird. The reason of this is that the edges of the new feathers are not of the same colour as the inner parts. Only the margins show, because the feathers of a bird overlap like slates on a roof, or the scales of a fish. After a time the edges of the new feathers become worn away, and then the differently-hued deeper parts begin to show, so that the bird gradually resumes the appearance it had before the moult. When the redstarts reach India in September most of the cocks are grey birds, because of the grey margins to their feathers; by the middle of April, when they begin to depart, many of them are black, the grey margins of the feathers having completely disappeared; other individuals are still grey because the margins of the feathers are broader or have not worn so much.

October is the month in which the falconer sallies forth to secure the hawks which will be employed in "the sport of kings" during the cold weather. There are several methods of catching birds of prey, as indeed there are of capturing almost every bird and beast. The amount of poaching that goes on in this country is appalling, and, unless determined efforts are made to check it, there is every prospect of the splendid fauna of India being ruined. The sportsman is bound by all manner of restrictions, but the poacher is allowed to work his wicked will on the birds and beasts of the country, almost without let or hindrance.

The apparatus usually employed for the capture of the peregrine, the shahin and other falcons is a well-limed piece of cane, about the length of the expanse of a falcon's wings. To the middle of this a dove, of which the eyelids have been sewn up, is tied. When a wild falcon appears on the scene the bird-catcher throws into the air the cane with the luckless dove attached to it. The dove flies about aimlessly, being unable to see, and is promptly pounced upon by the falcon, whose wings strike the limed cane and become stuck to it; then falcon and dove fall together to the ground, where they are secured by the bird-catcher.

Another method largely resorted to is to tether a myna, or other small bird, to a peg driven into the ground, and to stretch before this a net, about three feet broad and six long, kept upright by means of two sticks inserted in the ground. Sooner or later a bird of prey will catch sight of the tethered bird, stoop to it, and become entangled in the net.

A third device is to catch a buzzard and tie together some of the flight feathers of the wing, so that it can fly only with difficulty and cannot go far before it falls exhausted to the ground. To the feet of the bird of which the powers of flight have been thus curtailed a bundle of feathers is tied. Among the feathers several horsehair nooses are set. When a bird of prey, of the kind on which the falconer has designs, is seen the buzzard is thrown into the air. It flaps along heavily, and is immediately observed by the falcon, which thinks that the buzzard is carrying some heavy quarry in its talons. Now, the buzzard is a weakling among the raptores and all the other birds of prey despise it. Accordingly, the falcon, unmindful of the proverb which says that honesty is the best policy, swoops down on the buzzard with intent to commit larceny, and becomes entangled in the nooses. Then both buzzard and falcon fall to the ground, struggling violently. All that the bird-catcher has to do now is to walk up and secure his prize.

October marks the beginning of a lull in the nesting activities of birds, a lull that lasts until February. As we have seen, the nesting season of the birds that breed in the rains ends in September, nevertheless a few belated crow-pheasants, sarus cranes and weaver-birds are often to be found in October still busy with nestlings, or even with eggs; the latter usually prove to be addled, and this explains the late sitting of the parent. October, however, is the month in which the nesting season of the black-necked storks (*Xenorhynchus asiaticus*) begins, if the monsoon has been a normal one and the rains have continued until after the middle of September. This bird begins to nest shortly after the monsoon rains have ceased. Hard-set eggs have been taken in the beginning of September and as late as 27th December. Most eggs are laid during the month of October. The nest is a large saucer-shaped platform of twigs and sticks. Hume once found one "fully six feet long and three broad." The nest is usually lined with grass or some soft material and is built high up in a tree. The normal number of eggs is four, these are of a dirty white hue.

NOVEMBER

It is the very carnival of nature,
The loveliest season that the year can show!

The gently sighing breezes, as they blow,
Have more than vernal softness. . . .

BERNARD BARTON.

The climate of Northern India is one of extremes. Six months ago European residents were seeking in vain suitable epithets of disapprobation to apply to the weather; to-day they are trying to discover appropriate words to describe the charm of November. It is indeed strange that no poet has yet sung the praises of the perfect climate of the present month.

The cold weather of Northern India is not like any of the English seasons. Expressed in terms of the British climate it is a dry summer, warmest at the beginning and the end, in which the birds have forgotten to nest.

The delights of the Indian winter are enhanced for the Englishman by the knowledge that, while he lives beneath a cloudless sky and enjoys genial sunshine, his fellow-men in England dwell under leaden clouds and endure days of fog, and mist, and rain, and sleet, and snow. In England the fields are bare and the trees devoid of leaves; in India the countryside wears a summer aspect.

The sowings of the spring cereals are complete by the fifteenth of November; those of the tobacco, poppy and potato continue throughout the month. By the beginning of December most of the fields are covered by an emerald carpet.

The picking of the cotton begins in the latter part of October, with the result that November is a month of hard toil for the ponies that have to carry the heavy loads of cotton from the fields into the larger towns. By the middle of the month all the *san* has been cut and the water-nuts have been gathered in. Then the pressing of the sugar-cane begins in earnest. The little presses that for eight months have been idle are once again brought into use, and, from mid-November until the end of January, the patient village oxen work them, tramping in circles almost without interruption throughout the short hours of daylight.

The custard-apples are ripening; the cork trees are white with pendent jasmine-like flowers, and the loquat trees—the happy hunting ground of flocks of blithe little white-eyes—put forth their inconspicuous but strongly scented blossoms. Gay chrysanthemums are the most conspicuous feature of the garden. The shesham and the silk-cotton trees are fast losing their leaves, but all the other trees are covered with foliage.

The birds revel, like man, in the perfect conditions afforded by the Indian winter; indeed, the fowls of the air are affected by climate to a greater extent than man is.

Those that winter in England suffer considerable hardship and privation, while those that spend the cold weather in India enjoy life to the uttermost.

Consider the birds, how they fare on a winter's day in England when there is a foot of snow lying on the ground and the keen east wind whistles through the branches of the trees. In the lee of brick walls, hayricks and thick hedges groups of disconsolate birds stand, seeking some shelter from the piercing wind. The hawthorn berries have all been eaten. Insect food there is none; it is only in the summer time that the comfortable hum of insects is heard in England. Thus the ordinary food supply of the fowls of the air is greatly restricted, and scores of field-fares and other birds die of starvation. The snow-covered lawn in front of every house, of which the inmates are in the habit of feeding the birds, is the resort of many feathered things. Along with the robins and sparrows—habitual recipients of the alms of man—are blackbirds, thrushes, tits, starlings, chaffinches, rooks, jackdaws and others, which in fair weather avoid, or scorn to notice, man. These have become tamed by the cold, and, they stand on the snow, cold, forlorn and half-starved—a miserable company of supplicants for food. Throughout the short cold winter days scarcely a bird note is heard; the fowls of the air are in no mood for song.

Contrast the behaviour of the birds on a winter's day in India. In every garden scores of them lead a joyful existence. Little flocks of minivets display their painted wings as they flit hither and thither, hunting insects on the leaves of trees. Amid the foliage warblers, wood-shrikes, bulbuls, tree-pies, orioles and white-eyes busily seek for food. Pied and golden-backed woodpeckers, companies of nuthatches, and, here and there, a wryneck move about on the trunks and branches, looking into every cranny for insects. King-crows, bee-eaters, fantail and grey-headed flycatchers seek their quarry on the wing, making frequent sallies into the open from their leafy bowers. Butcher-birds, rollers and white-breasted kingfishers secure their victims on the ground, dropping on to them silently from their watchtowers. Magpie-robins, Indian robins, redstarts and tailor-birds likewise capture their prey on the ground, but, instead of waiting patiently for it to come to them, they hop about fussily in quest of it. Bright sunbirds flit from bloom to bloom, now hovering in the air on rapidly-vibrating wings, now dipping their slender curved bills into the calyces.

On the lawn wagtails run nimbly in search of tiny insects, hoopoes probe the earth for grubs, mynas strut about, in company with king-crows and starlings, seeking for grasshoppers.

Overhead, swifts and swallows dash joyously to and fro, feasting on the minute flying things that are found in the air even on the coolest days. Above them, kites wheel and utter plaintive cries. Higher still, vultures soar in grim silence. Flocks of emerald paroquets fly past—as swift as arrows shot from bows—seeking grain or fruit.

In the shady parts of the garden crow-pheasants look for snakes and other crawling things, seven sisters rummage among the fallen leaves for insects, and rose-finches pick from off the ground the tiny seeds on which they feed.

The fields and open plains swarm with larks, pipits, finch-larks, lapwings, plovers, quail, buntings, mynas, crows, harriers, buzzards, kestrels, and a score of other birds.

But it is at the *jhils* that bird life seems most abundant. On some tanks as many as sixty different kinds of winged things may be counted. There are the birds that swim in the deep water—the ducks, teal, dabchicks, cormorants and snake-birds; the birds that run about on the floating leaves of water-lilies and other aquatic plants—the jacanas, water-pheasants and wagtails; the birds that wade in the shallow water and feed on frogs or creatures that lurk hidden in the mud—the herons, paddy-birds, storks, cranes, pelicans, whimbrels, curlews, ibises and spoonbills; the birds that live among sedges and reeds—the snipe, reed-warblers, purple coots and water-rails. Then there are the birds that fly overhead—the great kite-like ospreys that frequently check their flight to drop into the water with a big splash, in order to secure a fish; the kingfishers that dive so neatly as barely to disturb the smooth surface of the lake when they enter and leave it; the graceful terns that pick their food off the face of the *jhil*; the swifts and swallows that feed on the insects which always hover over still water.

Go where we will, be it to the sun-steeped garden, the shady mango grove, the dusty road, the grassy plain, the fallow field, or among the growing crops, there do we find bird life in abundance and food in plenty to support it.

This is not the breeding season, therefore the bird choir is not at its best, nevertheless the feathered folk everywhere proclaim the pleasure of existence by making a joyful noise. From the crowded *jhil* emanate the sweet twittering of the wagtails, the clanging call of the geese, the sibilant note of the whistling teal, the curious *a-onk* of the brahminy ducks, the mewing of the jacanas and the quacking of many kinds of ducks. Everywhere in the fields and the groves are heard the cawing of the crows, the wailing of the kites, the cooing of the doves, the twittering of the sparrows, the crooning of the white-eyes, the fluting of the wood-shrikes, the tinkling of the bulbuls, the chattering of the mynas, the screaming of the green parrots, the golden-backed woodpeckers and the white-breasted kingfishers, the mingled harmony and discord of the tree-pies, the sharp monosyllabic notes of the various warblers, the melody of the sunbirds and the flycatchers. The green barbets also call spasmodically throughout the month, chiefly in the early morning and the late afternoon, but the only note uttered by the coppersmith is a soft *wow*. The hoopoe emits occasionally a spasmodic *uk-uk-uk*.

The migrating birds continue to pour into India during the earlier part of November. The geese are the last to arrive, they begin to come before the close of October, and, from the second week of November onwards, V-shaped flocks of these fine birds may be seen or heard overhead at any hour of the day or night.

The nesting activities of the fowls of the air are at their lowest ebb in November. Some thirty species are known to rear up young in the present month as opposed to five hundred in May. In the United Provinces the only nest which the ornithologist can be sure of finding is that of the white-backed vulture.

Some of the amadavats are still nesting. Most of the eggs laid by these birds in the rains yielded young ones in September, but it often happens that the brood does not emerge from the eggs until the end of October, with the result that in the earlier part of the present month parties of baby amadavats are to be seen enjoying the first days of their aerial existence. A few black-necked storks do not lay until November; thus there is always the chance of coming upon an incubating stork in the present month. Here and there a grey partridge's nest containing eggs may be found. As has been said, the nesting season of this species is not well-defined.

The quaint little thick-billed mites known as white-throated munias (*Munia malabarica*) are also very irregular as to their nesting habits. Their eggs have been taken in every month of the year except June.

In some places Indian sand-martins are busy at their nests, but the breeding season of the majority of these birds does not begin until January.

Pallas's fishing-eagle is another species of which the eggs are likely to be found in the present month. If a pair of these birds have a nest they betray the fact to the world by the unmusical clamour they make from sunrise to sunset.

The nesting season of the tawny eagle or wokab (*Aquila vindhiana*) begins in November. The nest is a typical raptorial one, being a large platform of sticks. It may attain a length of three feet and it is usually as broad as it is long; it is about six inches in depth. It is generally lined with leaves, sometimes with straw or grass and a few feathers. It is placed at the summit of a tree. Two eggs are usually laid. These are dirty white, more or less speckled with brown. The young ones are at first covered with white down; in this respect they resemble baby birds of prey of other species. The man who attempts to take the eggs or young of this eagle must be prepared to ward off the attack of the female, who, as is usual among birds of prey, is larger, bolder and more powerful than the male. At Lahore the writer saw a tawny eagle stoop at a man who had climbed a tree and secured the eagle's eggs. She seized his turban and flew off with it, having inflicted a scratch on his head. For the recovery of his turban the egg-lifter had to thank a pair of kites that attacked the eagle and caused her to drop that article while defending herself from their onslaught.

DECEMBER

> Striped squirrels raced; the mynas perked and pricked,
> The seven sisters chattered in the thorn,
> The pied fish-tiger hung above the pool,
> The egrets stalked among the buffaloes,
> The kites sailed circles in the golden air;
> About the painted temple peacocks flew.
>
> ARNOLD. *The Light of Asia.*

In the eyes of the Englishman December in Northern India is a month of halcyon days, of days dedicated to sport under perfect climatic conditions, of bright sparkling days spent at the duck tank, at the snipe *jhil*, in the *sal* forest, or among the Siwaliks, days on which office files rest in peace, and the gun, the rifle and the rod are made to justify their existence. Most Indians, unfortunately, hold a different opinion of December. These love not the cool wind that sweeps across the plains. To them the rapid fall of temperature at sunset is apt to spell pneumonia.

The average villager is a hot-weather organism. He is content with thin cotton clothing which he wears year in year out, whether the mercury in the thermometer stand at 115° or 32°. However, many of the better-educated Indians have learned from Englishmen how to protect themselves against cold; we may therefore look forward to the time when even the poorest Indian will be able to enjoy the health-bringing, bracing climate of the present month.

By the 1st December the last of the spring crops has been sown, most of the cotton has been picked, and the husbandmen are busy cutting and pressing the sugar-cane and irrigating the poppy and the *rabi* cereals.

The crop-sown area is covered with a garment that, seen from a little distance, appears to be made of emerald velvet. Its greenness is intensified by contrast with the dried-up grass on the grazing lands. In many places the mustard crop has begun to flower; the bright yellow blooms serve to enliven the somewhat monotonous landscape. In the garden the chrysanthemums and the loquat trees are still in flower; the poinsettias put forth their showy scarlet bracts and the roses and violets begin to produce their fragrant flowers.

The bird choir is composed of comparatively few voices. Of the seasonal choristers the grey-headed flycatchers are most often heard. The fantail flycatchers occasionally sing their cheerful lay, but at this season they more often emit a plaintive call, as if they were complaining of the cold.

Some of the sunbirds are still in undress plumage; a few have not yet come into song, these give vent only to harsh scolding notes. From the thicket emanate sharp sounds—*tick-tick*, *chee-chee*, *chuck-chuck*, *chiff-chaff*; these are the calls of the various warblers that winter with us. Above the open grass-land the Indian skylarks are singing at Heaven's gate; these birds avoid towns and groves and gardens, in consequence their song is apt to be overlooked by human beings. Very occasionally the oriole utters a disconsolate-sounding *tew*; he is a truly tropical bird; it is only when the sun flames overhead out of a brazen sky that he emits his liquid notes. Here and there a hoopoe, more vigorous than his fellows, croons softly—*uk*, *uk*, *uk*. The coppersmith now and then gives forth his winter note—a subdued *wow*; this is heard chiefly at the sunset hour.

The green barbet calls spasmodically throughout December, but, as a rule, only in the afternoon. Towards the end of the month some of the nuthatches and the robins begin to tune up. On cloudy days the king-crows utter the soft calls that are usually associated with the rainy season.

December, like November, although climatically very pleasant, is a month in which the activities of the feathered folk are at a comparatively low ebb. The cold, however, sends to India thousands of immigrants. Most of these spend the whole winter in the plains of India. Of such are the redstart, the grey-headed flycatcher, the snipe and the majority of the game birds. Besides these regular migrants there are many species which spend a few days or weeks in the plains, leaving the Himalayas when the weather there becomes very inclement. Thus the ornithologist in the plains of Northern India lives in a state of expectancy from

November to January. Every time he walks in the fields he hopes to see some uncommon winter visitor. It may be a small-billed mountain thrush, a blue rock-thrush, a wall-creeper, a black bulbul, a flycatcher-warbler, a green-backed tit, a verditer flycatcher, a black-throated or a grey-winged ouzel, a dark-grey bush-chat, a pine-bunting, a Himalayan whistling thrush, or even a white-capped redstart. Indeed, there is scarcely a species which inhabits the lower ranges of the Himalayas that may not be driven to the plains by a heavy fall of snow on the mountains. Naturally it is in the districts nearest the hills that most of these rare birds are seen—but there is no part of Northern India in which they may not occur.

The nesting activity of birds in Upper India attains its zenith in May, and then declines until it reaches its nadir in November. With December it begins again to increase.

Of those birds whose nests were described last month the white-backed vulture, Pallas's fishing-eagle, the tawny eagle, the sand-martin and the black-necked stork are likely to be found with eggs or young in the present month.

December marks the beginning of the nesting season for three large owls—the brown fish-owl, the rock horned-owl and the dusky horned-owl. The brown fish-owl (*Ketupa ceylonensis*) is a bird almost as large as a kite. It has bright orange orbs and long, pointed aigrettes. Its legs are devoid of feathers. According to Blanford it has a dismal cry like *haw, haw, haw, ho*. "Eha" describes the call as a ghostly hoot—a *hoo hoo hoo*, far-reaching, but coming from nowhere in particular. These two descriptions do not seem to agree. There is nothing unusual in this.

The descriptions of the calls of the nocturnal birds of prey given by India ornithologists are notoriously unsatisfactory. This is perhaps not surprising when we consider the wealth of bird life in this country. It is no easy matter to ascertain the perpetrators of the various sounds of the night, and, when the naturalist has succeeded in fixing the author of any call, he finds himself confronted with the difficult task of describing the sound in question. Bearing in mind the way in which human interjections baffle the average writer, we cannot be surprised at the poor success that crowns the endeavours of the naturalist to syllabise bird notes.

As regards the call of the brown fish-owl the writer has been trying for the past three or four years to determine by observation which of the many nocturnal noises are to be ascribed to this species. With this object he kept one of these owls captive for several weeks; the bird steadfastly refused to utter a sound. One hoot would have purchased its liberty; but the bird would not pay the price: it sulked and hissed. The bird in question, although called a fish-owl, does not live chiefly on fish. Like others of its kind it feeds on birds, rats and mice. Hume found in the nest of this species two quails, a pigeon, a dove and a myna, each with the head, neck and breast eaten away, but with the wings, back, feet and tail remaining almost intact. "Eha" has seen the bird stoop on a hare. The individual kept by the writer throve on raw meat. This owl is probably called the fish-owl because it lives near rivers and tanks and invariably nests in the vicinity of water. The nest may be in a tree or on a ledge in a cliff. Sometimes the bird utilises the deserted cradle of a fishing-eagle or vulture. The structure which the bird itself builds is composed of sticks and feathers and, occasionally, a few dead leaves. Two white eggs are laid. The breeding season lasts from December to March.

The rock horned-owl (*Bubo bengalensis*) is of the same size as the fish-owl, and, like the latter, has aigrettes and orange-yellow orbs, but its legs are feathered to the toes. This owl feeds on snakes, rats, mice, birds, lizards, crabs, and even large insects. "A loud dissyllabic hoot" is perhaps as good a description of its call as can be given in words. This species breeds from December to April. March is the month in which the eggs are most likely to be found. The nesting site is usually a ledge on some cliff overhanging water. A hollow is scooped out in the ledge, and, on the bare earth, four white eggs are laid.

The dusky horned-owl (*Bubo coromandus*) may be distinguished from the rock-horned species by the paler, greyer plumage, and by the fact that its eyes are deep yellow, rather than orange. Its cry has been described as *wo, wo, wo, wo-o-o*. The writer would rather represent it as *ur-r-r, ur-r-r, ur-r-r-r-r*—a low grunting sound not unlike the call of the red turtle-dove. This owl is very partial to crows. Mr. Cripps once found fifteen heads of young crows in a nest belonging to one of these birds. December and January are the months in which to look for the nest, which is a platform of sticks placed in a fork of a large tree. Two eggs are laid.

The breeding season for Bonelli's eagle (*Hieraetus fasciatus*) begins in December. The eyrie of this fine bird is described in the calendar for January.

In the Punjab many ravens build their nests during the present month.

Throughout January, February and the early part of March ravens' nests containing eggs or young are likely to be seen.

Ordinarily the nesting season of the common kite (*Milvus govinda*) does not begin until February, but as the eggs of this bird have been taken as early as the 29th December, mention of it must be made in the calendar for the present month. A similar remark applies to the hoopoe (*Upupa indica*).

Doves nest in December, as they do in every other month.

Occasionally a colony of cliff-swallows (*Hirundo flavicolla*) takes time by the forelock and begins to build one of its honeycomb-like congeries of nests in December. This species was dealt with in the calendar for February.

Blue rock-pigeons mostly nest at the beginning of the hot weather. Hume, however, states that some of these birds breed as early as Christmas Day. Mr. P. G. S. O'Connor records the finding of a nest even earlier than that. The nest in question was in a weir of a canal. The weir was pierced by five round holes, each about nine inches in diameter. Through four of these the water was rushing, but the fifth was blocked by debris, and on this a pair of pigeons had placed their nest.

GLOSSARY

Arhar.   A leguminous crop plant which attains a height of four feet or more.

Chik.   A curtain composed of a number of very thin strips of wood. Chiks are hung in front of doors and windows in India with the object of keeping out insects, but not air.

Holi.   A Hindu festival.

Jhil.   A lake or any natural depression which is filled with rain-water at all or in certain seasons.

Kharif.   Autumn. Rice and other crops which are reaped in autumn are called *kharif* crops. Crops such as wheat which are cut in spring are called *rabi* crops. Two crops (sometimes three) are raised in India annually.

Megas.   Sugar-cane from which the juice has been extracted.

Rabi.   Spring. See *Kharif*.

Shikari.   One who goes hunting or shooting.

Tope.   A term applied to a grove of mango trees, artificially planted. Thousands of such topes exist in Northern India. In some places they are quite a feature of the landscape.

Printed in January 2023
by Rotomail Italia S.p.A., Vignate (MI) - Italy